MORE ADVANCE PRAISE FOR
THE JAZZ OF PHYSICS

"Whether he's hanging with Brian Eno or Brian Greene, Alexander never loses sight of the math or the melodies, never condescends to his reader, but rather uses his own childlike awe and personal charm to take us into the details of chords and equations. It's impossible to resist following him as he 'solos with the equations of D-branes' on paper napkins in jazz clubs, searching for the eloquent underlying harmonies that brought the universe (and us) into being."
 —K. C. Cole, author of *Something Incredibly Wonderful Happens* and
  The Universe and the Teacup: The Mathematics of Truth and Beauty

"Music, physics, and mathematics have lived in tune since Pythagoras and Kepler, but Stephon Alexander's book creates a new and powerful resonance, coupling the improvisational world of jazz to the volatile personality of quantum mechanics, and making the frontiers of cosmology and quantum gravity reverberate like in no other book."
 —João Magueijo, professor of physics, Imperial College London,
  and author of *Faster Than the Speed of Light*

"*The Jazz of Physics* is a cornucopia of music, string theory, and cosmology. Stephon Alexander takes his reader on a journey through hip hop, jazz, to new ideas in our understanding of the first moment of the big bang. It is a book filled with passion, joy, and insight."
 —David Spergel, Charles Young Professor of Astronomy and department
  chair, Department of Astrophysical Sciences, Princeton University

THE JAZZ OF PHYSICS

Eric,
Thought of you as soon as
its publication was announced.
Love, Dad
June 2016

THE SECRET LINK BETWEEN
MUSIC AND THE STRUCTURE
OF THE UNIVERSE

THE
JAZZ
OF
PHYSICS

STEPHON ALEXANDER

BASIC BOOKS
A Member of Perseus Books Group
New York

Published by Basic Books,
A Member of the Perseus Books Group

Books published by Basic Books are available at special discounts for bulk purchases in the United States by corporations, institutions, and other organizations. For more information, please contact the Special Markets Department at the Perseus Books Group, 2300 Chestnut Street, Suite 200, Philadelphia, PA 19103, or call (800) 810-4145, ext. 5000, or e-mail special.markets@perseusbooks.com.

Set in 11.5-point Adobe Garamond Pro

Names: Alexander, Stephon, author.
Title: The jazz of physics : the secret link between music and the structure
    of the universe / Stephon Alexander.
Description: New York : Basic Books, [2016] | Includes bibliographical
    references and index.
Identifiers: LCCN 2015050500| ISBN 9780465034994 (hardcover) | ISBN
    9780465098507 (ebook)
Subjects: LCSH: Cosmology. | Universe. | Special relativity (Physics) | Space
    and time. | Musicology. | Music—Acoustics and physics. |
    Music—Philosophy and aesthetics.
Classification: LCC QB981 .A54 2016 | DDC 523.1—dc23
LC record available at http://lccn.loc.gov/2015050500

10 9 8 7 6 5 4 3 2 1

Dedicated to my parents, Felician and Keith,
and my daughter Kolka

CONTENTS

Introduction   1

1   Giant Steps   11

2   Lessons from Leon   27

3   All Rivers Lead to Cosmic Structure   41

4   Beauty on Trial   51

5   Pythagorean Dream   69

6   Eno, the Sound Cosmologist   85

7   Thriving on a Riff   93

8   The Ubiquity of Vibration   101

9   The Defiant Physicists   117

10   The Space We Live In   125

11   Sonic Black Hole   137

12   The Harmony of Cosmic Structure   145

13   A Journey into Mark Turner's Quantum Brain   159

14   Feynman's Jazz Pattern   171

15   Cosmic Resonance   179

16 The Beauty of Noise   189

17 The Musical Universe   203

18 Interstellar Space   215

    Epilogue   229

    Acknowledgments   *233*
    Notes   *235*
    Index   *243*
    About the Author   *255*

INTRODUCTION

It occurred to me by intuition, and music was the driving force behind that intuition. My discovery was the result of musical perception.

 —Albert Einstein (when asked about his theory of relativity)

And I cherish more than anything else the Analogies, my most trustworthy masters. They know all the secrets of Nature, and they ought least to be neglected in Geometry.

 —Johannes Kepler

Humans are special. Eleven *billion* years after the birth of the universe, conditions were just right for the seething, mineral-rich oceans of the planet we call Earth to spawn life—a mutating, evolving, hungry survivor. In just the last sliver of the life of the universe, we humans have grown both to cultivate Earth and unabashedly look to the skies to understand where we come from.

The people of every culture have pondered their origins and the origins of the cosmos. What is this space around us? Where did we come from? It is no mistake that these questions—questions that many of us asked as children—remain some of the most pressing in science. Questions like these point both to our innate curiosity about our origins and, as questions do, to the limits of our knowledge. For millennia we could only answer these questions with myth. But since the scientific revolution, we have tried to put myth aside, leaving the exploration of human and universal origins to scientists and their hard-fact methodologies. Modern cosmologists, though armed with fancy equations and high-tech experiments, can be said to be the myth makers of our time. Despite our precision mathematics and experiments, new surprises in modern physics and cosmology have emerged that compel some of the most able physicists to resort to myth making to try and explain the mind-bending information they have uncovered about the nature of the universe.

Heroic attempts have been made to relate the concepts that underlie modern cosmology to lay readers, but it is all too easy for books to fall short of the promise to explain. Putting into words subjects like general relativity and quantum mechanics, which are naturally communicated with the language of mathematics, is a tall order. These complex equations can blind even physicists, who themselves are challenged to completely understand or visualize what their formulas are telling them, a fact that emphasizes the need to find other ways to conceptualize the structure of the universe through clear physical pictures or analogies. I've found that the most successful books in communicating ideas are those that find the best analogies to mirror the physics. In fact, analogical reasoning will be a key driving force behind this book.

This book will take you on a firsthand journey into the process of discovery in theoretical physics research. We will see that, contrary to the logical structure innate in physical law, in our attempts to reveal new vistas in our understanding, we often must embrace an irrational, illogical process, sometimes fraught with mistakes and improvisational thinking. Although it is important for both jazz musicians and physicists to strive for technical and theoretical mastery in their respective disciplines, innovation demands that they go beyond the skill sets they have mastered. Key to innovation in theoretical physics is the power of analogical reasoning. In this book, I will show how the art of finding the right analogies can help us break new ground and traverse the hidden quantum world to the vast superstructure of our universe.

In this book, music will be the analogy that not only will help us understand much of modern physics and cosmology but will help unveil some recent mysteries that physicists face. Even while writing this book, this analogical thinking enabled me to discover a new approach to a long-standing unresolved problem in early universe cosmology. One of the biggest questions in understanding this, and a huge open issue in cosmology, is how the first structures emerged from an empty and featureless infant universe. The intricate way that the fundamental laws of physics work together to create and sustain the overarching structure of the universe, responsible for our very existence, seems like magic—not unlike how the bare bones of music theory have given rise to everything from "Twinkle Twinkle Little Star" to Coltrane's *Interstellar Space*. By using an interdisciplinary focus, inspired by three great minds (John Coltrane, Albert Einstein, and Pythagoras), we can begin to see that the "magical" behavior of the blossoming cosmos is based in music.

About a decade ago, I sat alone in a dim café on the main drag of Amherst, Massachusetts, preparing for a physics faculty job presentation when an urge hit me. I found a pay phone with a local phone book and mustered up the courage to call Yusef Lateef, a legendary jazz musician, who had recently retired from the music department of the University of Massachusetts, Amherst. I had something I had to tell him.

Like an addict after a fix, my fingers raced through the pages anxiously seeking the number. I found it. The brisk wind of a New England autumn hit my face as I called him. At the risk of rudely imposing, I let the phone ring for quite a while.

"Hello?" a male voice finally answered.

"Hi, is Professor Lateef available?" I asked.

"Professor Lateef is not here," said the voice, flatly.

"Could I leave him a message about the diagram that John Coltrane gave him as a birthday gift in '61? I think I figured out what it means."

There was a long pause. "Professor Lateef is here."

We spoke for nearly two hours about the diagram that appeared in his acclaimed book *Repository of Scales and Melodic Patterns,* which is a compilation of a myriad of scales from Europe, Asia, Africa, and all over the world.[1] I expressed how I thought the diagram was related to another and seemingly unrelated field of study—quantum gravity—a grand theory intended to unify quantum mechanics with Einstein's theory of general relativity. What I had realized, I told Lateef, was that the same geometric principle that motivated Einstein's theory was reflected in Coltrane's diagram. Einstein was a hero of mine. So were Coltrane and Lateef.

Professor Lateef shared some important information with me about how the diagram approximated cycles of fourths and fifths. He also had a deep interest in philosophy and physics and schooled me on his concept of autophysiopsychic music, which is music from one's physical, mental, and spiritual self.[2] This concept would have a major impact on my subsequent research relating music and the cosmos. Lateef encouraged me and vindicated my hope that there was a deep connection between music and the structure of the universe. That day, like a stereoscopic image coming into focus, my parallel lives in physics and jazz blended before my eyes, creating a new dimension.

Coltrane was fascinated by Einstein and his ideas. Einstein is famous for what is perhaps his greatest gift: the ability to transcend mathematical limitations with physical intuition. He would improvise using what he called *gedankenexperiments* (German for thought experiments), which provided him with a mental picture of the outcome of experiments no one could perform. For example, Einstein imagined what it would feel like to ride a beam of light. It takes insight to do this successfully. Another resource for Einstein was music. Though not a well-known fact, Einstein played the piano. Elsa, his second wife, once remarked: "Music helps him when he is thinking about his theories. He goes to his study, comes back, strikes a few chords on the piano, jots something down, returns to his study." On the one hand, Einstein used mathematical rigor; on the other, creativity and intuition. He was an *improviser* at heart, just like *his* hero, Mozart. Einstein once said, "Mozart's music is so pure and beautiful that I see it as a reflection of the inner beauty of the universe."

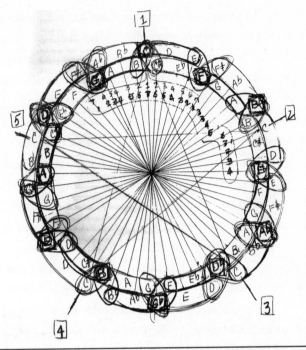

FIGURE INT.1. Diagram that John Coltrane gave to Yusef Lateef as a birthday present in 1961. Any other reproduction of the image is prohibited. *Ayesha Lateef.*

What the Coltrane mandala made me realize was that improvisation is a characteristic of both music and physics. Much like Einstein working with his thought experiments, some jazz improvisers construct mental patterns and shapes when they solo. I suspect that this was true of Coltrane.

John Coltrane passed away in 1967, two years after the cosmic microwave background radiation, a relic of the big bang itself, was discovered by Arno Penzias and Robert Wilson. The discovery crushed the theory of a static universe and confirmed an expanding one, as predicted by Einstein's theory of gravity. Among Coltrane's last recorded albums were three entitled *Stellar Regions, Interstellar Space,* and *Cosmic Sound.* Coltrane played with physics in his music and, incredibly, correctly realized that cosmic expansion is a form of antigravity. In jazz combos, the "gravitational" pull comes from the bass and drums in the rhythm section. The songs in *Interstellar Space* are a majestic display of Coltrane's solos expanding, freeing themselves from the gravitational pull of the rhythm section. He was a musical innovator, with physics at his fingertips. Einstein was an innovator in physics, with music at his fingertips. Nevertheless, what they were doing was not new. They were both *reenacting* the connection between music and physics, which had been established thousands of years earlier when Pythagoras—the Coltrane of his time—first worked out the mathematics of music. Pythagoras's philosophy became "all is number," and music and the cosmos were both manifestations of this philosophy. In the mathematics of the orbits of the planets rang "the music of the spheres," playing a harmony with the tones of a vibrating string.

Following in Coltrane and Einstein's steps, in this book we, too, will revisit the ancient realm where music, physics, and the cosmos were one. We will see how Pythagoras and others began to understand sound, and how their ideas and practices, in the minds of great thinkers like Kepler and Newton, developed into our present-day understanding of the dynamics of strings and waves. Twenty-five hundred years later, the inventors of string theory are busy investigating how fundamental strings might be used to unify the four forces of nature. Yet how many of them remember or place importance on the fact that one of the central equations of their theory, the wave equation, is rooted in a search for the universal connection to music?

This book is also an exercise in the power of *analogies.* By reconnecting the disciplines of physics and music through analogy, we can begin to understand physics through sound. We will see that harmony and resonance are universal phenomena and can be used to explain the dynamics of the early universe. As we'll see, a host of cosmological data reveals that roughly fourteen billion years ago, a relatively simple set of sound patterns developed into structures such as galaxies and galaxy clusters, which eventually enabled the development of planets and life itself.

We will also consider life's quantum origins. In most music, the range of tones in a scale is limited to discrete vibrations. The subatomic realm is also made up of discrete packets, known as quanta (hence quantum mechanics). After the Large Hadron Collider's recent confirmation of the Higgs boson particle, we have verification that the paradigm that underlies much of physical reality is quantum field theory. It's a mathematically daunting area of physics. Fortunately, for us, much of it can be understood in terms of the elements of music. For instance, broken quantum symmetries are vital in producing our fundamental forces and particles, just the way broken symmetries in musical structures, such as a major scale, create a sense of compositional resolution. Improvisation will play in our exploration, providing us with a tool for understanding the quirky dynamics of the quantum world: its inherent uncertainties and the idea that each outcome is actually the sum of all possible outcomes.

Next to mathematics, I learned that one of the most powerful tools involved with unraveling the secrets in the theoretical sciences is simplifying the system at hand and borrowing an analogy from what might, at first glance, be a completely unrelated discipline. It is in the *limits* of these analogies, where there exists a need for further research, that an avenue for discovery lies. It's like interdisciplinary rock hopping, from the bank of ignorance to the bank of knowledge, while the wide river of life rushes past.

While physics has reached unprecedented success in unveiling nature's secrets, from the smallest and largest imaginable distances, it's no secret that the physics community is in a state of crisis. Physicists are stuck with fundamental problems, like the apparent "fine-tuning" of the universe, an example being the delicate balance of the relative strengths of the four forces in nature that forms carbon-based life. I want to encourage the idea that physics can enter a new era of being inclusive and interdisciplinary—an improvisational physics. Rooted in cross-disciplinary analogies, improvisational physics pushes boundaries at the limits of analogy.

This is my journey. I was an adolescent, the son of a New York cab driver from Trinidad, when I rebelliously fell in love with a book called *The*

Privilege of Being a Physicist by Victor Weisskopf. My family expected music to be the object of my desire. "There are two things that make life worth living," said the author and Nobel laureate Victor Weisskopf, "Mozart and quantum mechanics." I loved Mozart but didn't know about quantum mechanics. This turned out to be the beginning of a long love affair that would become my future and would include more than quantum mechanics and Mozart: cosmology and John Coltrane would be the heart of this driving passion. But becoming a physicist has taken me places even those names couldn't have predicted. I have paved an unconventional path fusing jazz and physics to arrive at my profession as a theoretical physicist. It has only been possible because, over the last twenty years, I have been mentored and empowered by many teachers and friends, such as the Nobel laureate and superconductivity and music-loving pioneer Leon Cooper, as well as physics-enthusiast musicians, such as Ornette Coleman and Brian Eno. They taught me the importance of interdisciplinary thought and of seeing how analogies could be used to push the boundaries of knowledge.

Meeting these influential figures is part of the journey. Tapping into the beats and grooves of music theory is part of the journey. Tracing the evolution of structure in our universe is part of the journey. Creating an analogy between physics and music is part of the journey. *Not* having an adequate analogy, and needing rigorous calculations for clarity, is part of the journey.

How to Read This Book

This book will explore a lot of modern physics, relativistic cosmology, and music theory, but it requires no background in those fields; everything is self-contained. I have found over the years that learning through storytelling is an engaging and effective device for transmitting complex ideas in physics, and there are many stories in this book that contain profound concepts. From time to time, there will be some beautiful equations, but it's not necessary to understand these equations to grasp a concept. If you encounter an equation that you don't quite get, I encourage you to skip it and continue reading. Usually I find from my own experience that after a qualitative grasp of a concept, I can understand the equation in hindsight. Nonetheless, the equations are usually derived and explained in words.

I invite you, the reader, to a ball where physics and music dance with each other. I invite you, the reader, to research and question with me. I invite you, the reader, to take our musical analogies seriously and ask if we can learn something new about the universe through them. Let's improvise.

1

GIANT STEPS

Music is the pleasure the human mind experiences from
counting without being aware that it is counting.
                                        —Gottfried Leibniz

On a sunny summer day, Ruby Farley—Mum to her grandchildren—
sat sternly in her rocking chair, a flowery Caribbean head tie wrapped
around her head. Children were playing Wiffle ball outside her Bronx
brownstone. With her melodious Trinidadian accent, Mum cried, "Ah
don't care if yuh have to sit down and practice on de piano for hours,
yuh not leaving until yuh learn dat song!" Her eight-year-old grandson found it hard to place his fingers correctly on the keys. He was on
the verge of tears because the only music he could hear were the joyful
sounds of his friends playing outside. Suddenly, Mum's grimace soft-
ened. She smiled and sang to herself, "Ah can see it now, my grandson's
name in lights on Broadway." Mum had saved her money working as a
nurse's aide in the Bronx for thirty years in hopes of getting me there,
but I never became that concert pianist.

Ruby Farley, my father's mother, grew up in Trinidad in the forties,
when Trinidad was a British colony, and moved to New York City in
the sixties. There was a dynamic exchange of music between the Caribbean and New York at the time, and Mum carried more music with

her than just her Trinidadian accent. When she would return to New York from Trinidad, she would bring records of calypso greats, like the Mighty Sparrow and Lord Kitchener, and it was through those albums that I was exposed to the merging of soul music with calypso indigenous to Trinidad. This "soul of calypso," or soca music, was a fusion of the East Indian and African cultures. The music was developed in the late sixties and reached its modern form, which included influences of soul, disco, and funk music, in the seventies, when Trinidadian artists were recording in New York City.

To Mum, and many Afro-Caribbeans of her generation, being a musician was one of the few professions that afforded economic and social mobility. Mum's grand plans for me to study classical music and become a concert pianist began even before my parents and siblings left Trinidad to stay with Mum for a few years when I was eight. It was her way of handing me a ticket to the economic freedom that she and my parents had never tasted. My piano teacher Mrs. Di Dario, an Italian woman in her seventies, was a strict taskmaster. For an average eight-year-old, five years of learning études and memorizing scales with her was arduous, but it was the implicit pressure of the need to succeed that made it oppressive. Still, while I didn't enjoy the tedium of practicing under Mum's expectant eye, the classical composers, whose music I was learning to play, hooked my adolescent curiosity. They were able to put the scales together to make music! The idea that so many songs could come from only twelve keys fascinated and absorbed me. As I practiced, I became distracted by thoughts that were shaping into profound questions. How did people come to invent this thing called *music*? Why was it that when I played a major scale, I felt happy? C, E, and G, all notes in the key of C major, were happy notes. These are the first three notes in the first line of Elvis Presley's song "I Can't Help Falling in Love"—"Wise (C) men (G) say (E)." But when my finger climbed down from that white E key to the black E-flat key, that happy sound became sad. *Why?*

I was more interested in how music worked than learning to play others' compositions. This intrigue stayed with me throughout my youth and adulthood, but it couldn't focus my attention on regular practice. Eventually, my melodious Trinidadian grandmother realized that her money wasn't enough to nurture talent and so put up the white flag, and the piano lessons ended.

At that time I was attending Public School 16 and was in Mrs. Handler's third-grade class. My handwriting was bad, and my shy and inquisitive nature was interpreted as "slow." I had a near miss being put in a class for the mentally challenged because teachers doubted what my parents didn't. But I was spared, and one day found myself on a life-changing field trip. Back in those days, there were programs for public school classes to go to Broadway plays and museums for free. Mrs. Handler's class got to go see the dinosaurs at the Museum of Natural History. We students walked in a line holding hands through the grand corridors of stuffed creatures, seemingly ready to pounce, sleep, feast, or cry out.

Approaching the main hall, I noticed a smaller corridor to the left. At the back of the line, like a curious cat, both determined and naïve enough to potentially sacrifice one of its nine lives, I took the liberty of sneaking off to see what was there. I found myself alone staring at some papers protected by a thick glass pane. The writing on them was hiero-glyphic and clearly handwritten. To my eight-year-old eyes, it didn't look like it was from this planet. Then I saw the picture of the person behind these puzzles. His wiry hair formed an uncombed white-gray halo. His intensely focused gaze was calm but touched with mischief. I pictured him hunched over a desk, scribbling in code, perhaps hum-ming to himself in satisfaction or grunting in frustration. It was the first time I had seen Albert Einstein and his equations describing his theory of relativity. The magic began.

I didn't know that those enchanting scribbles described time and space as a single, changeable entity. But I did feel that those moments of contemplation stretched into what felt timeless. My eyes darted back and forth between Einstein's image and the symbols he had written. I sensed I was like him, and not just because my curly Afro resembled his wild locks, but because I saw a loner who liked to play with sym-bols and ideas the way I liked to play with musical notes on paper to make my own songs and try to answer my own questions. I wanted to know more. I wanted to figure out what he wrote. Something in me felt that, whatever Einstein was, I wanted to become like him. In those moments, I realized that there was something beyond my reality in Mrs. Handler's third-grade class, beyond the Bronx, maybe beyond this world, and that it had to do with those enigmatic symbols written down long ago by Albert Einstein, now preserved behind glass.

Fast-forward four years. In the early eighties in the Bronx, most teenagers, including me, were consumed with hip-hop, a music that reflected our experiences and backgrounds. It blended the funk of James Brown and Parliament with the extemporaneous lyrical forms of Caribbean and Latin music. A handful of my neighborhood friends would go on to become successful hip-hop producers and artists. My friend Randy, who later became Vinny Idol, was the most memorable. Randy was a tall, handsome twelve-year-old music fanatic of Jamaican ancestry who lived in a building on my newspaper delivery route. We bonded over our shared love for and understanding of music. I would frequently stop at Randy's apartment, and he would play me soul music from his record collection. He would often jam on his electric bass to the records. *That,* I liked. There it was again—that fascination in me to use notes to create, not just reproduce. Improvisation. This was my first true taste of it.

I had a room in the attic of our home that became my "mad scientist" lab—my workshop for experimentation. It was filled with disassembled radios, failed electronic toy projects, and a Marvel comic-book collection. Most every night, before going to sleep, I would play the radio stations that were popular with seventh-graders—98.7 Kiss FM or 107.5 WBLS—but one night, I decided to look for a new station. As I turned the tuning knob, half-hoping to find a new beat to share with my friends, my ears autofocussed onto a sound that, at first, I mistook for the white noise one hears between stations. It wasn't. After a few seconds, I recognized it as a saxophone. The music initially seemed chaotic and random, but it filled me with a mysterious energy that kept my attention locked on the station. I was spellbound by this sound, and I listened until the song's completion. The disk jockey came on and said, "You were just hearing the free jazz music of Ornette Coleman." There it was again. Improvisation.

My father was a big fan of the saxophone and noticed my growing interest in it. He and my mother got me a used vintage alto sax that they purchased at a garage sale from the wife of New York Mets baseball player Tim Teufel. My parents paid fifty dollars for the sax, and despite the few dings and the worn-out lacquer, it played well. I later joined the band at my middle school, John Philip Sousa Junior High School, which Mr. Paul Piteo, a professional jazz trumpeter, led. He showed me how to get notes out of the sax and to make my own reeds. "Finally," I thought to myself, "I don't have to practice." Armed with tools of musical independence, I could play free jazz, like my friend Randy, like Ornette Coleman. I could just improvise. *This* was fun. *This* was music to me. Nothing like practicing the piano.

Looking back today, I had no idea how wrong I was. I had fun mimicking and jamming to the popular songs on my portable radio, but there's no free lunch in free jazz. In traditional jazz, there are well-defined melodic themes and harmonic movements throughout a song. During my early days as a student of jazz, I used to think that playing free jazz meant that anyone can pick up an instrument with no training or practice and improvise meaningfully. As I matured musically and came to understand the rules of harmony and the basic forms in the standard jazz tradition (which I will later discuss), I discovered that free jazz does have its own internal structure and is an extension of the standard jazz tradition. A free jazz musician has very little structure to depend on and is challenged to improvise something that moves the audience. But what is this thing we call music?

Music is deeply human.[1] Everyone has different tastes and preferences in music. I have some friends who only listen to electronic music; others think that jazz is the only thing worth listening to. I also know people who believe the only "real" music is classical. There is a growing community of individuals that enjoy noise music. Given the difficulty in finding a definition of music that works for everyone, I am going to restrict our discussion of music to the classical Western tradition. I do this because much of the music discussed in this book is based on the classical Western twelve-tone system. In general, a piece of music can be represented as a complex sound waveform that evolves in time. Contained within this waveform are perceived elements such as tone, meter, rhythm, pitch, melody, and harmony.[2]

Defining the various elements of Western music is a subtle matter. For the sake of brevity, I will provide a simplified description. Imagine hearing a song that begins with just a piano; that discrete sound is an example of *tone* or *note*. A tone can be perceived to have a definite frequency (or pitch) that belongs to a specific musical scale with a finite set of frequencies. A *melody* is a succession of tones that is usually the main theme in a musical piece. We all have a favorite melody—mine is "My Favorite Things." Dancers pay special attention to *meter*, which is the recurring coherent pattern of accents that provide the pulse or beat and is important in developing the song's *rhythm*. Beats in a meter are grouped in bars. For example, the meter in the waltz will have beats in groups of three for each bar, while a techno beat will have four recurring beats. *Harmony* involves the consonant or dissonant relationship between notes played simultaneously, and these chords create a movement between musical tension and release.

Music is a physical event, and like most nontrivial physical systems, it has structure or, as musicians call it, *form*. Just as the skeleton determines the shape of an animal, the musical form provides the scaffold for melody, rhythm, and harmony to unfold in a coherent manner. In many cases, at the beginning of a composition, a motif or theme is initiated. We see this in a lot of classical and baroque music. One of the most famous motifs is the first four notes in Beethoven's *Fifth Symphony*: ta ta ta taaaa. This motif can be tied together into a *phrase*, which is the musical equivalent of a sentence, a grouping of notes with a coherent musical sense.

Phrases can fit into a given *chord*, or *key*. In many popular forms of music, the chords will change and eventually return back to the home

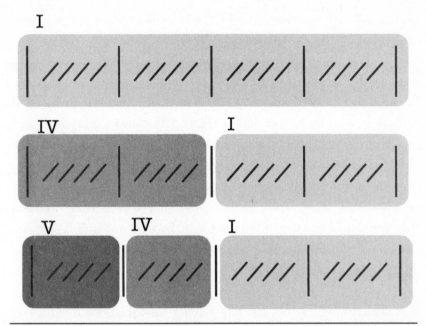

FIGURE 1.1. Schematic of the twelve-bar blues structure. The meter is usually four beats per bar. In the key of C, the form begins with the tonic: the first note of the scale (I) repeats itself for four bars then ascends to the fourth (IV), which is F. In the last four bars, the harmony eventually resolves to the tonic.

chord. Many songs begin with a home key and take a journey deviating away from that key with a return to the home key, usually designated by the Roman numeral I. A typical chord movement in most Western music is the II-V-I progression. In the key of C, this corresponds to D-G-C. One of my favorite II-V-I songs is Cole Porter's "Night and Day," made popular by Frank Sinatra. Another common music form is the blues, which uses twelve bars that move from I–IV with a few repetitions and an eventual return to I. Listen to any B.B. King song and you can hear how this progression functions.

These forms create a progression: tensions and resolutions are created, tapping into human feeling and the aspect of storytelling. Our description of music started with a single note, then chords, phrases, rhythms, and forms—a complex structure that started with a wave and its characteristic wavelength and frequency. All of this opens the doors to human emotion and creativity—using notes to express the personal and connect the self to nature. It emerges almost as if by magic. Indeed, music is deeply human.

Although most popular songs from rock, pop, and jazz are based on straightforward forms like that in Figure 1.1, modern composers, such as György Ligetti, based some of their compositional structures on more intricate self-similar forms like fractals. In these forms, the smaller piece mirrors the form of the larger structure. Many structures in nature, such as snowflakes, leaves, and coastlines, have fractal properties.[3] Research has found a fractal structure in some of Bach's compositions.[4] Fractal structures in music occurs when shorter musical lines are mirrored in longer musical passages.

For me, playing my sax was like playing basketball. I did it for fun. It was a hobby, a passion of the times. But lurking somewhere in the depths of myself was that tickle to know more, not just about how to create music but also about its greater origins, its link to our emotions, how to get that something called "music" from "notes." What *were* "notes" anyway? What I hadn't realized yet is that science would help give me those answers. Science would become my true passion.

John Philip Sousa Junior High was situated in the Edenwald Projects, off Baychester Avenue. A few years before I matriculated to Sousa, it was rated as one of the most crime-ridden and dangerous middle schools in the country. That was until Dr. Hill Brindle came on board. Brindle, as we called him, was a towering militant and possessed an articulate baritone voice. His fatherly presence evoked a mixture of admiration, respect, and fear in both the students of the school and the thugs that lurked in the neighborhood. Sousa was a public school, but Brindle ran it like a military private school. As a student at West Point military academy, Brindle had been an Olympic hopeful for the 400-meter dash. But while training on the track one day, Brindle was shot in his thigh by an unknown gunman, destroying his Olympic dreams. His energies needed a new focus, so maybe that was the reason why he later joined Dr. Martin Luther King's civil rights movement and ended up devoting himself to the education of inner-city students. Every morning at Sousa, Brindle and his deans would be at the school's two entrances to check that every individual came in with his or her notebook and textbooks. And every Wednesday, Brindle himself delivered a gospel-like speech to the entire school, all students expected in semiformal attire.

During one significant and fortuitous Wednesday morning assembly, Brindle, composed as ever, informed us that we had a special guest, and then exited the stage. I was in eighth grade, and I have never forgotten that day. An older man in an orange jumpsuit with a boom box on his shoulder appeared from behind the stage's curtains. A well-known hip-hop beat was thumping out of the radio. Some students started laughing as if the man were a clown; others were confused. But most enthusiastically bobbed their heads to the music. The guy certainly got our attention. Then he shut off the music and introduced himself. He was Fredrick Gregory, an African American astronaut. Gregory, speaking with his native Washington, DC, accent, asked, "How many y'all liked that beat? Wasn't it cool?" The students cheered, smiles on their faces. "Yeah . . . that beat was cool!" he said. The assembly had become a party. Then the astronaut asked, "But do you know how this radio was able to make this music?" He continued, "It's powerful to own a radio and hear this music, but true power is in the ability to make one of these radios. Knowing how this radio works helped me to become an astronaut. I studied science. I went to college and I got an engineering degree." It was a powerful message. *He* had come to the school, to *us*, and to make it clear why science was important. And he was like us, culturally, socially, economically, geographically. He put it simply. "I'm from the same background as y'all, and if I could do it, so can you!" Science. It wasn't the first time I had thought about studying science. But this was different.

However, I didn't begin my journey in science to become a physicist, describing the physics of music or unraveling Einstein's equations; I wanted to become a roboticist. Next to my radio, my secondhand sax, and all my cluttered bedroom experiments sat my stack of Marvel comics. Tony Stark, the comic-book superhero who *built* his own Iron Man suit, was a huge inspiration. After that school assembly, although I kept playing my sax in the middle-school jazz band, my main fascination became science.

One day shortly before graduation, Mr. Piteo pulled me aside and said, "Son, you are one of the two most talented music students I ever had. The other guy is the director of the Apollo Theater band. I can get you into the High School of Performing Arts. No problem." Being able to attend the top music high school in New York City was a tremendous opportunity and would have made Mum proud. But I never told her about it because I had other ideas. I was on the science path and chose to attend DeWitt Clinton High School.

My first day at DeWitt Clinton, which had around six thousand students, threw me. I was sitting in my English class, discussing Hamlet, when the class was disturbed by the sounds of young men rhyming with each other. Outside the classroom windows, a sea of Latino students were playing handball, break dancing, and engaging in "free-style battle rap." This was an improvisational form of rhyming with rhythmic complexity, and battle rappers would compete with each other while being judged by enthusiastic onlookers. Miss Bambrick, our jolly Irish Shakespearian teacher, interrupted our lesson with an enthusiastic "Now *that's* mastery of the English language!"

Life took a turn. I would cut the classes I found boring and take the bus to the basketball courts. It was on the courts that we would both play pickup ball, taking breaks to rap and break-dance on flattened refrigerator cardboard boxes. On the bus, I was joined by others from my high school, who were also cutting class. I would overhear the discussions of some guys who called themselves Five Percenters. They'd debate about humanoid-like aliens that came from space to interact with "the original Asiatic black man." No joke. I overheard tidbits of other sci-fi topics and realized they really believed them! My high school was one of the meccas of the Five Percenter Nation, and no one messed with them. Those guys were far from feeble and seldom smiled. I thought of the Five Percenters as another gang, but I turned out to be wrong about that. They were highly disciplined and committed to their spiritual and independent intellectual studies, and we had something in common— and it wasn't just that we were cutting class and toying with "science." A common practice for Five Percenters was "dropping knowledge," which is similar to an intellectual debate, sometimes in the form of battle rapping. We both wanted an escape from the dismal prospect the future held for us. I sought my escape through comic books, video games, and my newfound love for science. These guys adopted a worldview from their leader Clarence 13X, a former student of Malcolm X, who, after attaining spiritual enlightenment, spread the following gospel throughout the streets of New York City:

- 85 percent of the masses blindly follow religion.
- 10 percent of the masses are deliberately misleading the masses.
- 5 percent are enlightened and realize that they are "gods" of their own destiny.
- Mathematics is the language of reality, and in order to master nature, a Five Percenter must understand the mathematical patterns underlying nature: they called this supreme mathematics.

The Five Percenters were the 5 percent. So when the "gods" saw me repeatedly on the bus by myself, quietly playing with equations I was learning from my mathematics teacher, Mr. Daniel Feder, they would try to engage me in their debates about alien life-forms that had communicated with the original Asiatic black man. Eventually, they invited me to join them. Granted, I was fascinated with their speculations, but I refrained and proceeded with my precalculus homework. Although I never joined the Five Percenters, they admired and protected me from the thugs that often preyed on the weak or nerdy. And I admired them, because MC Rakim, a devout Five Percenter, had just released a debut album, *Eric B is President*, that had taken New York City and the world by storm. Rakim was, and still is, my favorite MC, and unlike today's hip-hop, his lyrics promoted self-knowledge and he approached his improvisational delivery like a scientist. Rakim has gone down in history as the greatest battle rapper because of his ingenious improvisational capabilities and the unique polyrhthmic cadence of his rap delivery. I sometimes like to think that the great mathematician Leibniz prophesied Rakim with his quote "music is the pleasure the human mind experiences from counting without being aware that it is counting." Rakim equates his rhymes to "dropping science." His hit song, "My Melody," went something like:

> *That's what I'm sayin', I drop science like a scientist.*
> *My melody's a code, the very next episode.*
> *Has the mic often distortin', ready to explode.*
> *I keep the mic at Fahrenheit; freeze MCs to make 'em colder.*
> *The Listener's system is kickin' like solar . . .*

I had my first physics class a year later when I was a sophomore. I was anxious. I wasn't alone. All the nerds who sat in the front of the room were also scared. A thin bespectacled man with a wild head of hair walked into the room and wrote a simple equation on the blackboard. Three characters and an equal sign: $F = ma$. Force equals mass times acceleration. An object will accelerate when an external force is applied to it. The more mass an object has the harder it is to accelerate with the same external force. We had never seen such an equation. Mr. Kaplan walked into the middle of the room, sat on an empty desk, and took a tennis ball out of his pocket. He threw the ball up in the air and caught it. He saw everyone was so attentive that he didn't have to do it again and, after a moment, asked, "What's the velocity of the ball when it returned back to my hand?" The class sat silent. No one knew what to say. And in that minute or two, magic began to happen. I pictured that tennis ball going up, coming to rest in the air above our heads, and landing again in Kaplan's hands. Again I saw it happen. And again. I became that ball. My hands trembled, and perhaps, somehow, my eyes deceived me. Kaplan honed in on me.

"What's your name?" he asked.

"Stephon," I said.

"So Stephon, what do you think?"

And to my amazement afterward, the words just spilled out: "The ball will have the same velocity as it did when it left your hand."

Kaplan had a *big* smile on his face "That's exactly the answer! This is a sacred principle of nature called the conservation of energy."

Kaplan proceeded to the blackboard and showed, through simple addition and multiplication and using the symbols of the equation, F, m, and a, how the energy was conserved. Here was a "sacred" principle, demonstrated and intuited, with a tennis ball. For the first time in my life, certain events came together and made sense. I understood something about the world in a way I hadn't before. The equations brought me back to my face-to-face with Einstein seven years earlier and the attraction I had felt to those mysterious hieroglyphs behind the glass. Here, the power of four symbols, precisely aligned, composed an equation that revealed the functioning of the ball. I came to learn they could describe almost any object in the world, even the planets in outer space. After class, Mr. Kaplan approached me and said, "The best physicists are blessed with the gift of intuition. You have that. Come to my office later." The Five Percenter influence in the back of my brain made me wonder if I was about to be inducted into a secret society.

Daniel Kaplan had been a trained master composer and a jazz baritone sax player and later was drafted to serve in the Korean War. During the war, he worked on radar technology. Consequently, Kaplan caught the physics bug and upon his return, pursued graduate studies in physics, while still playing his saxophone and composing.

He would be the person that would solidify my passion to become a physicist. Kaplan was the chair of *both* the music and science departments. When I walked into his office, there was a large picture of Albert Einstein and, across from it, another picture of jazz saxophone player John Coltrane. It was the first time I'd seen them together, and I wondered why Mr. Kaplan would have a picture of a jazz musician along with a physicist. Coltrane would later become my favorite jazz musician because of our shared admiration for Einstein. "You have great physical intuition, but to become a physicist, you need to learn a lot of mathematics. It's the language of physics," he said. I told him that I had read a little about Einstein—that matter can be converted to energy. He responded, and I'll never forget this, "You see that book?" pointing to a gigantic book called *Gravitation.*[5] "It's all about Einstein's general theory of relativity. It reveals the secrets behind space, time, and gravity. If you want to be a physicist you have to go to college. Then, when you go to graduate school, you can take general relativity." He continued, "Come into my office anytime you want to read these books, or if you have any questions, whatsoever."

I visited Kaplan's office every spare moment. I read his books and talked physics and music. I'd skip lunch. One day, Kaplan gave me an album, *Giant Steps* by John Coltrane. This groundbreaking album, released in 1960, was, in hindsight, a demonstration of Coltrane's "sheets of sound" and a sonic equivalent to Einstein's bending of the space-time fabric. I ended up joining my high-school jazz band and took calculus at the City College of New York, all under Kaplan's encouragement. And then almost everything started to shift.

During the mideighties, America transmuted from bell-bottoms to Spandex, from Jimmy Carter to Ronald Reagan, and the Bronx was bubbling with creativity in the arts. One of my best friends, Harvey Fergurson, knowing I played sax, invited me to join his new hip-hop band called Timbukk 3, which was under the mentorship of hip-hop pioneers Afrika Bambaataa and Jazzy Jay. Afrika Bambaataa is known for spreading hip-hop throughout the world and forming the Universal Zulu Nation, which used hip-hop culture to provide peaceful alternatives for gang members. Timbukk 3 was poised to be the Bronx wing of a collective of "conscious hip-hop artists" called Native Tongue. A Tribe Called Quest, The Jungle Brothers, and De La Soul were just a few noteworthy members of Native Tongue. Strong City Studios was Bambaataa's north Bronx recording studio. There, I sampled beats and had my sax sampled by Jazzy Jay. I loved the excitement of being in the recording booth. With my alto pointed to the mic, I watched as

Jazzy Jay and Harvey, bopping their heads by the soundboard, recorded the rhythmic modified Coltrane riffs they elicited from me, which they later chopped into sample bits and included throughout their rap song. Everything was so *on*: collaboration seemed to make creativity soar, and within a few months, Timbukk 3 was offered a recording deal. It was 1989. Hip-hop's international appeal and influence was exploding, and the doors were wide open for me to become a beat maker and producer. But, deep down, I knew that I needed to grow musically, especially on my sax. And deeper than that, physics tugged at me. The importance of equations and how things worked far outweighed the hip-hop path. And so I decided to attend college instead.

Growing up in the Bronx, despite its challenges and absurdities, was fertile ground for my becoming a physicist. Although my environment in the Bronx was filled with opportunities to become a working musician, the creative energy expressed by my peers—the battle rappers, break dancers, beat makers, and Five Percenters—and my devoted teachers (in particular Mr. Kaplan and Dr. Brindle) were able to inspire me to pursue my true passion. Having a role model like Mr. Kaplan, who showed me that I could identify as both a musician and a physicist, encouraged me to forge a career in physics. However, I would remain in conflict about whether or not I had made the right choice. To quiet my self-doubt, I knew I would have to figure out a way to have my physics and my music talk to each other. Lurking in the back of my mind was that image of Albert Einstein at one end of Mr. Kaplan's office and John Coltrane at the other end, representing the dialogue between physics and music that would be my life.

In college, things shifted again. I majored in physics. I pursued the professional path. I took a handful of courses on music theory, but the truth is, my involvement with music in college was minimal. It wasn't until my graduate school years that my search for a connection between music and physics was truly ignited.

2

LESSONS FROM LEON

Professor Leon Cooper was a fellow New Yorker and a Nobel Prize winner, coinventor of the Cooper pair. And there he stood at the head of my class, a physics genius in a suave Italian suit with perfectly groomed, wavy hair. The students in his advanced quantum mechanics class watched in awe as he improvised Feynman diagrams on the blackboard. Quantum mechanics describes our universe on the very smallest, subatomic scales, where the "stuff" in our universe takes on characteristics of both waves and particles. Energy and matter have a dual particle-like and wave-like nature. Get it? No? That's OK—neither do most physicists. It's an abstract and unintuitive theory, and the equations quickly get hairy and cumbersome. In 1948, Richard Feynman, an eventual Nobel Prize winner, sought a better way to think about quantum mechanics. The result was his simple and visually appealing Feynman diagrams, which completely changed how physicists handle complex particle interactions. Feynman was known for his vast array of interests, his genuine joie de vivre, and a teaching ethic based on simplifying concepts and delivering them clearly. He could make the most complicated ideas engaging for the youngest physics students. Feynman diagrams showed he could also engage the most established physicists, too.

Leon Cooper stood at the blackboard, a charismatic smile on his face, jotting down a diagram. The straight, wavy, and spiral lines, the

FIGURE 2.1. The large-scale structure of the universe. Each dot corresponds to a galaxy. Roughly one billion galaxies are pictured here. *Courtesy of Jamie Bock/ Caltech.*

forward and backward arrows, the symbols of a multitude of particles such as electrons, positrons, and quarks appeared and disappeared rapidly from the blackboard. Gradually, the Feynman diagram in Figure 2.2 emerged, which describes the quantum dynamics of the annihilation electron and its antiparticle. I could practically *see* the energy and matter, white as chalk, popping into and out of existence, morphing from one form to another, against the blackboard backdrop of our universe. Cooper was a master. Being able to do physics with him was like being a basketball fan who gets to play pickup with Michael Jordan.

In 1957, when Leon Cooper was twenty-seven, he, along with John Bardeen and Robert Schrieffer, successfully cracked a forty-year-old puzzle, explaining the quantum mechanical origin of a phenomenon called superconductivity, which won them a Nobel Prize.

The story of Cooper's Nobel began in 1911. The Dutch experimental physicist Heike Kamerlingh Onnes had discovered that when he

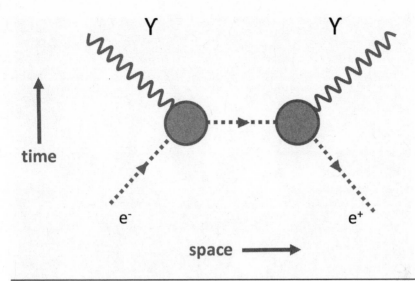

FIGURE 2.2. A Feynman diagram of an electron (e⁻) and an antielectron or positron (e⁺) annihilating each other to produce a photon of light (γ).

got a metal down to near absolute zero temperature (–452.11 degrees Fahrenheit)—an idealized lowest possible temperature at which a system has zero energy—electrons would flow through the metal without any resistance. This was a shocking observation. Electricity, as we find it in common circuitry, is the flow of electrons *against* a resistance that is inherent to the conductor. It works similarly to how friction slows a wheel on the surface of a road. Hence, a material with an *un*restricted flow of electrons is called a *super*conductor. It's as if the road suddenly changed so that the wheel could roll completely unimpeded on a perfectly smooth frictionless surface. Superconductors have since found many useful technological applications, many resting on the intimate relation between electric currents and the magnetic fields that are created whenever electric current is established.

After the emergence of quantum mechanics in the early 1900s, many great theorists, including Albert Einstein, tried to find the microscopic theoretical underpinnings of superconductivity but to no avail. A quantum mechanical picture of superconductivity did not exist. It was Cooper's ingenious physical insight, known as the Cooper pair,

FIGURE 2.3. Professor Leon Cooper, Brown University. *Courtesy AIP Emilio Segre Visual Archives, W. F. Meggers Gallery of Nobel Laureates.*

that unlocked the quantum secrets of superconductivity. Under typical circumstances, the individual electrons flowing in a piece of metal wire experience resistance because they repel each other, much like the defending players in rugby or football game interfere with the movement of the player with the ball. However, Cooper showed that by using the wave-like property of electrons, they can "pair up," thus changing their repulsive properties in the metal and allowing them to conduct without resistance.

Superconductivity, since it was quantized, meaning that only discrete bundles of energy exist as opposed to a continuous flow, has not waned in importance. Once Cooper, Bardeen, and Schrieffer figured out how superconductivity worked, others were able to find useful applications for it. Superconductivity is at the heart of magnetic resonance imaging (MRI)—medical scans used to probe anatomical form and function. Some situations, such as tests for tumors, require far more precision than an X-ray scan. Strong and uniform magnetic fields are needed. The efficient electric currents in a superconductor generate the large magnetic fields, which optimize the MRI scanning operation. Likewise, a superconducting quantum interference device, or SQUID, is used to detect incredibly weak magnetic fields. It is used in biology to measure, for instance, neural activity in the brain or other weak magnetic fields produced by minute physiological changes, such as in the heart of an unborn baby.

Magnetic train levitation is based on the Meissner effect, which is a direct result of superconductors. The influence between magnetic fields and electric currents implies that magnetic fields exert forces on electrons, which work to obstruct the flow of a current. Superconductors, with their unrestricted flow of electrons, will expel any magnetic field present in order to maintain the resistance-free flow of the current. Thus, if you place a magnet above a superconducting material, the material's supercurrent, which refuses to allow a magnetic field to enter it, produces a strong mirror image magnetic field, causing the external magnet to levitate. Train tracks made out of superconductors and train "wheels" made of magnets will induce the Meissner effect, causing the train to levitate. This effect played a key role in the discovery of the Higgs boson particle. The Higgs boson is nothing but a type of superconductivity, only now the superconducting medium is empty space itself. These achievements, which followed from the groundbreaking discovery of Cooper and his colleagues, are only a few of the reasons why Cooper was my hero.

By the end of high school, I had read Werner Heisenberg's books on matrix mechanics. Heisenberg is one of the inventors of quantum mechanics and is best known for the uncertainty principle, fundamental to the theory. I had also read *A Brief History of Time* by Stephen Hawking, a cosmologist famous for his work on the radiation emitted by black holes, and I tore through *Surely You're Joking, Mr. Feynman!*, excerpts from Feynman's multifaceted life. Reading everything I could get my hands on about physics provided a perfect escape while growing up in a part of the Bronx where reality, to many, was dismal.

I arrived at college armed with the passion to become a physicist, yet felt completely underprepared for the rigor of a physics major. I would spend hours staring at single pages, reading and rereading a paragraph or a set of equations until the concepts slowly seeped into my brain. I endured long exams and lab reports, while fueled on *lots* of coffee, my staple. Much of my undergraduate and early graduate years were spent feeling clueless and out of place, a dreadlocked Trinidadian from the Bronx.

Yet, I was willing to withstand years of self-doubt, my peers' underestimation, while I strove to gain competence as a researcher in physics. My undergraduate schooling trained me to manipulate equations used to describe the world around me. The ideas were fascinating, the difficulty understanding them discouraging. All along, deep down, I had this driving question: "Why is there something rather than nothing?" It was a nagging curiosity, like those days as a boy at the piano, when my attention would stray from the notes on the page to question the very existence of music and why it made me feel the way it did. Over the years, quantum theory slowly provided the keys to probe this fundamental question in physics.

And now, as luck would have it, two years into grad school, Leon, that spectacular quantum juggler, accepted me into his research group, the Cooper group, as his PhD student. I was shocked. It was a dream come true. As I got to know Leon, it became clear that he was a theoretical physicist who was neither limited nor defined by any subdiscipline. He had that Feynman-like quality of play and wonder in his pursuits. He taught courses with teachers from other disciplines with such enthusiasm that it was obvious that collaboration and the exchange of ideas was in his lifeblood. One of the most valuable lessons I learned from Leon was that carrying over concepts from one discipline to another is an art. By creating an analogy between a known idea in one field and an unresolved problem in another field, discoveries can be made and new avenues of exploration opened up. This showed up with the very first project that I undertook in the Cooper group.

Leon enjoyed working on interesting and seemingly insurmountable problems, regardless of the subdiscipline. He would fearlessly tackle the most difficult issues or even correct long-held misconceptions and paradigms in other fields such as radiation physiology, neuroscience, and philosophy. During the time I spent working in the Cooper group, Leon was working in the field of neuroscience. And so it was that I began life as a graduate student in physics studying the brain. Who would have thought computational neuroscience would turn me into a cosmologist?

Cooper was trying to construct a coherent theory of memory based on neural networks. A classic example of a neural network is the Hopfield model, which illustrates how associative memory works. Funnily enough, the basic idea behind the Hopfield model comes not from neuroscience but, surprisingly, from the quantum mechanical physics of magnetism, specifically, the Ising model, named after German physicist Ernst Ising.[1]

Consider a simplified picture of a magnet—a periodic, equally spaced array of atoms of a metal, such as iron. Every atom in that array will be defined by a quantity called spin. Quantum spin is much like that of a spinning top. But unlike the top, the electron's spin can have only one of two values: it either points up or down because the atomic spins are quantized. The atoms cannot take on a large number of arbitrary values of spin but only the two quantized, or discrete, values of up or down.

In this model, any charged particle, whether positively or negatively charged, can generate a magnetic field if it has spin. So far, that describes every individual atom. But when atoms combine in an organized group, new physics arises from their interactions. Some scientists call this an emergent phenomenon. If all of the atoms have spins pointing in the same direction, they will combine to generate a net magnetic field. Under normal circumstances, this is unlikely to occur because, at room temperature, the ambient thermal energy is enough to agitate the atoms to flip their spins so that they will have randomly oriented spin directions and will not exhibit any net magnetism.

The effect that one atom has on the other atoms around it is called interaction energy, a type of stored energy known as potential energy. Like all forms of potential energy in physics, this is a quantity that nature minimizes. For example, as you stretch a rubber band, its potential energy increases. Once you let it go, it will snap back to its original position, expending or converting that potential energy into kinetic energy or motion. This process minimizes the potential energy.

Feynman diagrams clarify quantum particle interactions. To help us understand complicated situations in physics, we call on another tool—mathematics. Mathematics is like a new sense, beyond our physical senses, that enables us to comprehend things that we cannot understand solely through our own perceptions or intuitions. In fact, many realms of physics, as well aspects of other sciences, like chemistry and biology, are highly counterintuitive. They follow rules that, although consistent and comprehensible, cannot be seen without using mathematics to extend our perceptions. In the case of atoms, spin, and magnetization, mathematics is used to clarify how complex systems of atoms behave. It is useful to first state the ideas intuitively, and then formalize the intuitions mathematically. Let's take a more mathematical look at the Ising model of magnetism used in Professor Cooper's research. We will devote some time to these details because many of these ideas will carry over for topics in the rest of the book.

The mathematics of spin will show how the interaction energy (E) works among atoms in the model. If the spin of one atom changes, we want to know how the spins of neighboring atoms are affected. Let's take an arbitrary atom, i. The atom i can take on positive integer values 1, 2, and so on, and thus specify an atom. For instance, $i = 1$ is atom 1, and $i = 3$ is atom 3. We can call the spin of atom i "S_i." Then $i = 1$ specifies S_1, the spin of atom 1, and so on. The spin of atom i's closest neighbor is "$S\_i + 1$." Now, $i = 1$ signals the spin of atom 1, S_1, *and* the spin, S_2, of its neighbor, atom 2.

When neighboring spins *agree*, i and $i + 1$ are either both up or both down. We can intuit that since they agree, then the interaction energy between them will be *less*. On the other hand, when neighboring

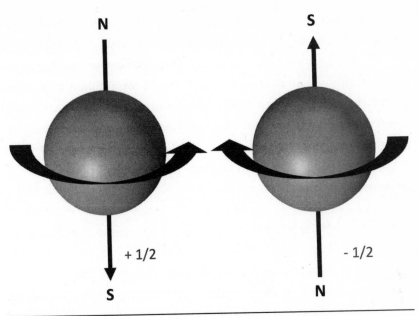

N

S

+ 1/2

S

- 1/2

N

FIGURE 2.4A. Quantum spin orientation for up and down states.

spins *disagree* such that, either spin S_i is up and its neighbor's spin S_i + 1 is down, or S_i is down and its neighbor S_i + 1 is up, then there is *more* interaction energy caused by the tension between the disagreeing spins. One can imagine two people in a discussion. If they agree, there will be less to discuss, less interaction. If they disagree, they will interact more, trying to shift the other's viewpoint.

Mathematically, if we treat an up spin as a plus one and a down spin as a minus one, we can then multiply S_i times S_i + 1 to get a plus one or minus one when *combining* the two spins. There are then only four possible outcomes: they are both up ($1 \times 1 = 1$) or both down ($-1 \times -1 = 1$), or the first is up and the second is down ($1 \times -1 = -1$), or the first is down and the next is up ($-1 \times 1 = -1$). For any pair of particles, if they are both in the same direction, then the product is S_i $\times$ S_i + 1 = 1, and if they are in opposite directions, it's S_i $\times$ S_i + 1 $= -1$. The single value of either plus one or minus one thus tells us whether the two atoms have the same spin or opposing spins. Two numbers reveal definitive physical information, and so we have our first mathematical representation of our atomic spin model.

In creating the Ising model of magnetism, we intuited that when neighboring spins agree, there will be less interaction energy, and when they disagree, there will be more interaction energy. But by how much the energy will increase or decrease for the entire array of atoms is yet to be determined. Before writing down the equation, let us review the elements of our model.

- The sample is a metal, for example iron, with all its atoms arranged in an array.
- Each atom in the sample has either spin up (1) or down (−1).
- If neighboring atoms have the same spin, the interaction energy (E) is low (with a numerical value of 1, because $1 \times 1 = 1$ and $-1 \times -1 = 1$).
- If neighbors have opposing spins, the interaction energy (E) is high (with a numerical value of −1, because $1 \times -1 = -1$ and $-1 \times 1 = -1$).

By summing up all the agreeing and disagreeing spins in our system, we can now write down the full equation describing how much interaction energy there is among the particles.

$$E = -J \sum_i S_i S_{i+1}$$

E is the interaction energy between particles, and it is equal to $S_i \times S_{i+1}$, which is 1 for agreeing spins and −1 for disagreeing spins. The sigma (Σ) symbol tells us to add up all these agreements or disagreements for all values of i, which generates the numeric value of how much agreement or disagreement there is overall. Note that if they all agree, that sum will be some large positive number; if they all disagree, it will be some large negative number. The J specifies how much

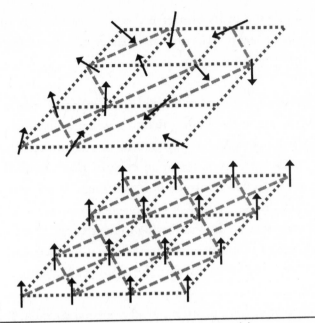

FIGURE 2.4B. Spin orientation S in the Ising model.

interaction energy there is depending on the sum of those agreements and disagreements. The larger the J is, the stronger the interaction energy between the spins. For instance, if J is 0.1, and if the sum of agreements happens to be 400, then the energy quantity will be 40. Finally, the minus sign shows that, all along, we treat agreements as decreasing the interaction energy and disagreements as increasing the interaction energy, so in this example, the interaction energy would be −40.

When the interaction energy is high due to disagreeing spins, the small magnetic fields of individual agreeing atoms are dominated, and the metal will *not* behave magnetically. But if the interaction energy is low because of lots of agreeing spins, the cumulative effect of the individual magnetic fields add up. Overall, the lowest energy configuration corresponds to spins that are predominantly aligned, and in that case, the substance will exhibit a magnetic field. It becomes a magnet. The potential energy has been minimized, which is equivalent to our rubber band being in its unstretched position. When atoms of iron align, the most highly magnetic material, it becomes a ferromagnet, and the mathematic model describing ferromagnetism is called the Ising model.

The important thing to take away from the Ising model is that it specifies the quantum physics dictating when a metal can become magnetic or not. Naturally, there are further conditions that can be imposed, such as the presence of an external magnetic field, but the aim here is not to delve into the inner workings of magnetism but to illustrate how the Ising model of ferromagnetism leads to the Hopfield model of neuroscience. It was amazing to me that the apparently unrelated Ising model of magnetism provided the direct inspiration for the Hopfield model of brain networks. The analogy was a beautiful one.

The Hopfield model is a classical model of neural circuitry. It took the Ising model of atomic spin interactions in a metal and reimagined the scenario as communicative interactions among neurons in a brain. The result was a neural network system that used extremely simple rules and mathematics to successfully store learned memories such that they could be retrieved or remembered. Loosely speaking, we can think of the shapes of local islands of correlated spins as the configurations that would be responsible for storing a memory.

Experiments reveal that neurons communicate with each other by "firing," or releasing neurotransmitters at junctions connecting them. These junctions are called synapses. John Hopfield simplified this complicated transmission between neurons by assigning a "strength" of the interaction between two neurons, equivalent to J in the Ising model. But in this model, the story is not so simple. In the Ising model, spins only interacted with their nearest neighbors, but within the intricacy of our brains, all neurons are connected to each other. The key analogy is between the spin in the Ising model and a neuron firing in the Hopfield model. If the value of the neuron is up, then the neuron has fired an electrochemical signal, and if the value of the neuron is down, then the neuron has not fired.

Hopfield proposed an equation for the overall "state" of the lattice of connected neurons, analogous to the overall energy in an array of interacting iron particles. The same mathematics that governs the Ising model also governs the Hopfield model: two neurons that both either fire, or do not fire, increase the state of connectedness, while neurons performing opposing actions decreases the state of connectedness. By simply replacing the spin variable, S ($S\_i,S\_i + 1, \ldots$), with the neuron variable, n ($n\_i, n\_i + 1, \ldots$), the models are *almost* identical.

$$E = \sum_{ij} w_{ij} n_i n_j$$

In place of a single J term for the value by which each pair changes the overall state, a separate term has been introduced with a *distinct* value for each pair, due to the interconnectedness of the neurons in the brain. The new variable in the model is the *weight*, $w_{i,j}$, between neurons i and j. It dictates how strongly a particular neuron, i, communicates with its neighbor, j—the efficacy of the synapse that connects these neurons. Setting these connection strengths turns out to be at the core of the Hopfield model. The array can fall into a large set of distinct states, mathematically dictated by the pattern of connection strengths throughout the array.

This analogy has limits: clearly human memory is more complex than a magnetized metal.[2] Humans are alive and can exert control. But neural networks based on these models can learn. My job in Cooper's lab was to work on unsupervised neural networks, an extension of these neural networks. While Hopfield's networks are *shown* what they are meant to learn so that they can train their synapses to identify patterns, unsupervised networks, by contrast, can train *themselves* to learn new memories. Typical applications of unsupervised neural networks occur when there is a great deal of data without any obvious preexisting categories. The network assumes an intelligence that identifies natural classes for the data to fall into. Satellite photos, stock trading patterns, and tweets on Twitter are all examples of large data pools that can be "data mined" via unsupervised learning. The Hopfield model is a classic example of carrying concepts from one field to another—and also explains why I was working in neuroscience while completing a physics degree.

I took away two important lessons working in the Cooper group. First, I will never forget the value and beauty of seeing and applying similar patterns from one field to another. Applying analogies across disparate fields is more of an art, I learned, than pure science. This also happens in music, when different music traditions are fused—for example when Coltrane took musical devices from other cultures and infused them into the jazz tradition. Coltrane skillfully integrated aspects of the Indian raga system into his improvisational repertoire,[3] which is especially interesting because there are some similarities among some Indian scales and modal jazz. This fusion occurs in one of Coltrane's most famous songs "My Favorite Things." Second, I realized that these analogies would always be limited, but it is their very limitations that provide the seeds for new insights and discoveries. In the case of the Hopfield model, thinking of a neuron as a quantum spin in a magnet was a useful analogy for giving neuroscientists the computational tools that physicists used to calculate the onset of magnetism. Of course the analogy was limited. Neurons are wired together in a much more complicated way than spins in a metal. But because that limitation was known, neuroscientists were able to focus on incorporating the complexity of the circuitry into the analogous model, and thus improve it. As a new researcher in Cooper's lab, I was in search of a new analogy for a special class of neural networks that I was investigating. I had no idea that it would come from outer space.

3

ALL RIVERS LEAD
TO COSMIC STRUCTURE

Providence, where I was attending graduate school at Brown University, was a small city in the smallest state in the United States. It was a shift from my upbringing in New York, but it had its scene, notably AS220 on Empire Street. I'd seek a hiatus from graduate school coursework and research and escape to this experimental jazz club in downtown Providence. There was a jazz band called The Fringe led by trombone great Hal Crook, whose free-flying trombone solos were a perfect blend of Ornette Coleman's free jazz and Crook's own mind-bending compositional algorithms. It was more than enough to inspire me to pick up my horn and begin a self-study of jazz. During the day I'd do my calculations, and during the night I'd play in jam sessions. These prepared me for the summers, when I'd return to New York and join in the sessions at Smalls Jazz Club in the West Village or travel over to Wally's Café Jazz Club in Boston. The sessions at Smalls were especially impactful as it wasn't too different from being in a science lab. Smalls, a cozy dive owned by Mitch, a former nurse and teacher, hosted the best musicians in New York City. Then there were the all-night jam sessions I attended, coupled with free lessons from legendary players. One of my teachers there was Sacha Perry who would show me alternative ways to solo over turnarounds.[1] Perry would often remark,

"Bud Powell showed us how to do this, but cats these days don't want to practice." Mrs. Di Dario was right after all.

During those six years in graduate school, my passion to play jazz saxophone grew to be commensurate with my passion for physics. With this dual interest, something new and powerful started to develop. I drew energy from playing with my newly formed jazz fusion band The Collective, not only while we played at AS220 and other urban gigs attended by jazz regulars, but to another kind of audience as well. Local campus coffee shop gigs became a natural avenue to draw an interdisciplinary crowd. Some audience members ignored the music; others even turned away in annoyance. But a good percentage of them got the vibe. The music would get their minds ticking about something new, something jazzy, something that, for me, often remained a mystery. Every individual in the band had a unique perspective to offer as I saw it. I realized there was a twofold drive in my playing: I was inspired to improvise, and I was inspired to create new connections based on feedback from members of the audience. Both were avenues for experimental thought.

It didn't hurt that my new mentor, cosmologist Robert Brandenberger, was a jazz lover. He encouraged me to pursue both my physics and jazz investigations by giving me complete freedom to formulate my own physics ideas, which usually occurred while I was listening to Hal Crook every Wednesday. I would bring my notebook to Hal's performances and improvise with equations and diagrams, while submerged in the ocean of Hal's intricate trombone riffs, members of the rhythm section spontaneously creating structure in their experimental jazz explorations. Whenever I would play, Robert would attend, bringing with him a stack of physics papers and ongoing calculations to read.

Robert's training was in constructive quantum field theory, and he had unprecedented knowledge of technical matters (like Monk did of music theory), such as differential geometry, the mathematics behind Einstein's theory of general relativity. But he did not limit himself by resorting solely to mathematics. Like Monk working with his angular melodic themes, Robert would compose his theories from ideas, no matter

FIGURE 3.1. Professor Robert Brandenberger. *Christina Buchmann.*

how strange they seemed when they first appeared. When he gathered with his students, it felt like an improvisation session at Smalls. He would engage in free-form group improvisation with his students. A student would present an idea. The idea may not have made any sense, but Robert would boomerang the idea back to the student in a more structured form. Students would learn from the master by playing with him in real time. By mimicking the master, we picked up the intuitive and technical skills needed to express our ideas more fully.

During this time I was taking Robert's course on general relativity: I was finally learning what those mysterious symbols behind the thick pane of glass written in Einstein's hand encoded and what secrets that huge book in Kaplan's office, *Gravitation*, held in its pages.

Robert was admired among grad students and was known as the professor that you could go to with any question whatsoever, no matter how seemingly idiotic, and he would turn it into something meaning-ful. Robert and I were both coffee addicts, and one day at our favorite coffee spot in Providence, Ocean Coffee, I asked Robert, "What is the most important question in cosmology?" I was probing for examples of data that might be amenable to study with the unsupervised learning mechanisms I was working on with Cooper. I expected to hear some-thing like, "What caused the big bang?" or "What are the fundamental building blocks of matter?" Robert was in a state of intense contem-plation for almost two minutes. I had time to observe him while he thought, and I remember him sitting with back upright, head bowed in a position that was slightly awkward yet perfectly meditative, his long thin hands resting on his knees. Wrist warmers filled the gap between his shirt cuffs and his hands, covering his bony wrists. And suddenly the answer was there. His eyes lifted quickly to meet mine, and he re-plied as if no time had elapsed for him whatsoever. "How did the large-scale structure in the universe emerge and evolve?" "*What*?!" I thought. But, I knew better than to hastily question him.

At the time, my education in physics had been limited to matters of our Earth, such as quantum physics and classical electrodynamics. Un-til that moment, it had never occurred to me that galaxies and super-clusters of galaxies were organized structures, let alone that they could tell us something profound about the nature of the universe, including what it is made of and how it came into being. In fact, I was even unaware that cosmologists studied these vast structures. I sat on his proposal for several weeks, and then it hit me. If there was a time in the universe's past when there was no structure—in the frenetic, seeth-ing conditions of the early universe, for instance—then understanding how its current structure came to exist and what caused it to organize itself would interconnect galaxies to stars, to planets, and ultimately, to human beings.

As far back as the second millennium BCE, astrologers sought patterns in the random distributions of stars in the night sky with a desire to find order and meaning in the cosmos. In Mesopotamia, China, Babylon,

Egypt, Greece, Rome, and Persia, an apparent order in our constellations was discovered. But there was more than met the eye. The first telescopes, built in the Netherlands in 1608 and quickly taken up and improved upon by Galileo Galilei in 1609, were able to detect light that humans couldn't see. They improved upon the human eye by magnifying and focusing light using lenses. What followed was an eight-hundred-year effort to build bigger telescopes capable of seeing more and more of our universe, culminating in part in Edwin Hubble's discovery that our universe had other galaxies. A typical galaxy is a pancake-like collection of hundreds of billions of stars, approximately ten thousand parsecs in diameter, rotating around a center, like a spinning Frisbee.

In 1920, a few years before Hubble's earth-shattering discovery, two leading astronomers, Harlow Shapley and Heber Curtis, were in a "great debate" about the scale of the universe. At this juncture in the history of cosmology, there was inconclusive evidence concerning the size of the universe. Astronomers were seeing some enigmatic spiral objects, which they called nebulae. According to Shapley, these nebulae were nothing more than spinning gas clouds contained within our galaxy. He believed that the universe was comprised of only one galaxy, no other galaxies existed outside our own. Curtis, however, argued that the nebulae were actually additional galaxies outside the Milky Way.

Hubble's discovery settled the debate, proving our universe had other galaxies, but what cosmologists at that time slightly overlooked was that the galaxies existed in clusters, and the extent of their clustering was on the order of one million parsecs, but this fact was not considered particularly interesting. Even after many years of mapping out the spatial distribution of galaxies over ever-larger distances, some astronomers remained uncertain that there was any interesting organization or larger clustering to the distribution of galaxies.[2]

Margaret Geller, however, thought differently. From a young age Geller was fascinated by patterns. When she was a child, her father, Seymour Geller, showed her the relationship between patterns in nature and physics. An X-ray crystallographer, he studied the relationship between the atomic structure of materials and their physical properties.

FIGURE 3.2. The Great Wall large-scale map discovered by Geller and Huchra. *Margaret Geller.*

When Margaret was at the Harvard-Smithsonian Center for Astrophysics, she deliberately sought out patterns in the large-scale distribution of galaxies, courageously probing outward with her telescope to unfathomable distances. In 1989, in a groundbreaking work with John Huchra, Margaret Geller mapped out galaxies extending distances on the order of one hundred million parsecs! They discovered galaxies that clustered into a filamentary wall-like structure that was popularly called the Great Wall—the largest observed structure in the universe.[3] This structure was the first indication that galaxies ordered themselves. But as Brandenberger had said to me, the question was how.

Encountering Geller's and Huchra's work for the first time, I suddenly saw the entire universe as a massive self-organizing network. This subject of large-scale structure also resonated with me because of what I had learned from biology—that three-dimensional structures often reveal the function of a biological system. An important example is the double helix structure of DNA, which informs the function of the coding of genomes and the interaction between proteins and DNA. Was this galactic large-scale structure showing us, in all its glory, why things work the way they do?

Cosmologists have used modern technology to help them find, like the ancients, order within the apparent randomness—and not just for the hundreds of millions of stars within our own galaxy but in the distribution of galaxies in our *entire* universe. A humble pursuit. They have dedicated themselves to searching for new galaxies and mapping

FIGURE 3.3. Cosmologist Margaret Geller. *Scott Kenyon.*

out their locations to investigate just how "correlated" galaxies are to each other.

In addition to creating and studying their cosmic maps, cosmologists have also considered the *dynamics* of galaxies. Highly massive galaxies attract each other because of their immense gravitational pull, thus affecting their motions through space. Cosmologists and astrophysicists have found that our universe is expanding: galaxies are moving apart from one another as our space-time expands, much like raisins in a loaf of bread move away from one another as the bread rises. The rate of expansion of the universe has changed throughout history and has played an essential role in the formation and evolution of galaxies—how they have taken shape and organized themselves into a large-scale structure. Indeed, this structure would not have formed if the universe had not expanded. This is a nontrivial fact. However, when cosmologists began to make sense of the previously unseen supergalactic structure, they could not agree on the nature of the structure itself. Some looked at the data and were convinced that these collections of galaxies were threaded in a filamentary structure, much like a spiderweb. Others argued that the space-time fabric of the universe was organized into bubble-like structures, with galaxies distributed on the surface of the bubbles. Determining what type of structure prevails in the cosmos is pivotal: it would tell us about the underlying physics that actually seeded the first stars, galaxies, and clusters of galaxies during the early stages of the universe.

These two arguments prompted Robert's answer to my question of what to study. Robert lamented the discord among cosmologists. I, for the first time, stewed over cosmic expansion and imagined the birth and growth of its largest structures as a massive self-organizing galactic network. Less as student than as a potential collaborator, I dreamed up a question for Robert: Why not train Leon's newly developed neural network on the large-scale structure data and let it decide what the true structure is? Leon Cooper's interdisciplinary approach proved contagious, and within a month, I was working on a joint project with Leon and Robert to use an unsupervised neural network to examine the large-scale structure. As the months progressed, I would continue asking Robert more and more questions. Robert was also an expert on Feynman diagrams. I was stuck on understanding the point where an electron and a positron annihilate each other and light suddenly gets created. I wondered if we were to zoom into that point we would require infinite energy, and I remained confused about its resolution.[4] One day Robert intuited the direction that all my questions were driving me in and said, "Ahhh . . . you want to find a quantum theory of gravity, don't you?" As we will see later, the unification of quantum mechanics and gravity becomes necessary at the shortest distances when elementary particles interact with each other, and it is considered to be one of the holy grails of fundamental physics research. By the end of the semester, I became Robert's PhD student. I began to work at the interface of quantum gravity and cosmology with the question of the origin and evolution of cosmic structure as a driving force in my research.

The neural network project was never completed for two reasons. Because of the tremendous data sets of galaxies, to make sense of them, the unsupervised networks required an extreme reliance on computer algorithms and coding. The more I started to study structure formation using Einstein's theory of relativity, the more I got hooked on relying on the power of a simple pen and paper to manipulate these beautiful equations over cups of coffee as opposed to writing thousands of lines of code on a supercomputer. Things are changing: today, in the world of physics theory, computers are becoming more central to research. Fortunately, programming is also becoming more fun. And it turns out that others later, independently, picked up on the idea to use neural networks to study large-scale structures, so maybe that project will be completed someday after all.

From quantum mechanics and magnets to neural networks to galactic clusters: I hadn't fully realized the adventure that had begun. I was to go on as a physicist, my head stuck in the cosmic fabric woven with galaxies. I would revisit quantum mechanics as I studied the subatomic sea of particles in the early universe that existed before galaxies, before planets, before humans, but at the time I had no idea. The connection was there, but I needed another stepping-stone, perhaps another analogy, to determine just how quantum physics could produce cosmic structure. It was going to take giant steps. After all, Robert targeted the problem as the greatest outstanding mystery in cosmology. Yes, this required giant steps.

"Giant Steps." The name of jazz saxophonist John Coltrane's renowned piece of improvisation, this piece of music uniquely investigated chord progressions and changed jazz forever. What would Coltrane, one of my greatest jazz idols, think of the universal structure technology has revealed today? He himself contemplated the cosmos and experimented with structures in his jazz compositions and improvisations. We'll get there. But first, the other side of the story. The story of *scientists* who have used *music* to explain the universe. Ancient philosophers such as Pythagoras, the first astrophysicists like Johannes Kepler—these remarkable mathematicians intuited that harmony and sound lie behind the creation of matter and the evolution of structure in the universe. Their theories paved the path for science as we know it today, but their analogies in music *did* fall short of reality—and they stopped there. But I didn't, and I am not alone in this quest.

4

BEAUTY ON TRIAL

During my last years as a graduate student, superstring theory was at its peak, and all theory graduate students that I knew were reading the latest papers by the great Edward Witten, the genius of the field. Like Paul Dirac, the English theoretical physicist had made fundamental contributions to quantum theory, Witten had world-class rigorous mathematical chops. He won a Fields Medal and possessed Einsteinian physical intuition. As soon as you thought you had gotten a handle on one of his research papers, a new groundbreaking paper would come out—a theory that had already gone through two revolutions with a third right around the corner. We would torture ourselves trying to keep up because string theory was exciting. It was simple in theory and complex mathematically, with a lot of room for new creative ideas.

String theory, at first, may seem counterintuitive. In classical physics, a vibrating physical string produces standing waves of integer frequencies. Zooming into the string reveals that it is made up of atoms. But according to string theory, if you zoom into an elementary particle, you will see a vibrating string of energy. In the string world, it's the strings that are fundamental. They only look like particles when you zoom out. String theory is even more musical than quantum field theory. Michio Kaku says it best: "The subatomic particles we see in nature, the quarks, the electrons are nothing but musical notes on a tiny vibrating string . . . Physics is nothing but the laws of harmony that you can write on vibrating strings . . . The universe is a symphony of vibrating strings."[1]

And then what is the mind of God that Albert Einstein eloquently wrote about for the last thirty years of his life? We now, for the first time in history, have a candidate for the mind of God. It is cosmic music.

Because of string theory's musical nature, I was able to quickly intuit the theory through my understanding of music and sound. Plus, it fit perfectly with my desire to combine my two passions. If the chord fits, play it. The various ways that a fundamental string can vibrate produces different tones, which translate into particle properties such as charge, mass, and spin. What's more, a particular vibration of the string gives the quantum of the gravitational field, the graviton. There it was, gravity emerging from quantum physics—finally. Like Einstein, many of the top minds in physics had tried but failed to unify quantum mechanics with gravitation, but a simple vibrating string elegantly did the job.

String theory naturally emerged as a theory of everything in the sense that the physics of this vibrating string gave back the particles of the force carriers and all the elementary particles. But this elegance came with a price. When string theorists started exploring the physics of the string, they found some surprises. A string moving in a small region of space experiences the same physics as a string moving in a large space: we call this target space duality (T-duality). String theory does not assume that we live in four dimensions; in fact the theory suggests that the world is ten dimensional. Even today, many clever particle-collider and space-based experiments are on the hunt to catch a glimpse of these extra, hidden dimensions. But in this rich new world of unification, it turned out that there was not a unique formulation of string theory. In fact there were five different string theories.

During my last year in graduate school, Brandenberger had a group meeting. He was one of the theorists in the early eighties who pushed the viability of using quantum field theory to address puzzles in the early universe. During this meeting, Robert told us that he felt that string theory had developed enough to begin to resolve some problems with early universe cosmology. I still remember him saying, "It will be great if we can find a string theory realization or alternative to inflation." It became clear to me what my calling would be over the next couple of years: I would learn enough string theory to figure out whether cosmic inflation could be a stringy phenomenon.

Right before leaving Brown, I walked into the office of string theorist Antal Jevicki, and behind closed doors, I asked him, "Antal, can you give me some advice about making it in my postdoc?" Antal with his Hungarian smile said, "Get on your feet and start running." The winners of the race would get a contract for a postdoctoral researcher position, until they landed a permanent paid position. As a postdoc, you are supposed to establish yourself as an independent researcher and make a big impact in your field. In theoretical physics, getting a postdoc is key to getting a faculty job. But it is not uncommon for a scientist to spend as long as a decade in postdoc purgatory before landing a job. Competition for these research positions is fierce, and the chance of landing a postdoc at a leading institution is highly unlikely. I found out later that out of over three hundred candidates who applied to the postdoc position at Imperial College in London, only two were chosen. Fortunately, due to some independent work I had done during my last year of graduate school, I was one of the chosen ones.

So I finally flew Robert Brandenberger's coop and went straight across the Atlantic to Imperial College, one of Europe's meccas for theoretical physics. At that juncture, I naïvely thought that Brandenberger's and Cooper's encouragement for outside-the-box thinking and improvisational methods in physics was common practice. Imperial would at first prove to be the opposite. The fear of failure and the feeling of being an imposter, which still lurked in my subconscious, sprang forth as a brand-new postdoctoral fellow. Although I was accustomed to feeling somewhat isolated from my peers, having been the only black physics PhD student at Brown (and one of three in the US), I now faced the reality of also being one of two Americans in the European network of theoretical physics postdocs. And it didn't help when I met the other postdocs, especially my office mate, Jussi Kalkkinen, a towering and monkish Finnish string theorist who would lock himself in his office for marathon calculation sessions, manipulating the equations of eleven-dimensional supergravity. Any attempt on my part to mimic this practice soon left me dozing off in my office after two hours of calculating.

Concerned about what it took to be a successful postdoc, I wondered what my own path would be in a field that seemed to be mostly about "shut up and calculate."[2] I quickly realized that I was in way over my head since the other postdocs possessed far better technical and mathematical chops than I had at the time. So what counts in theory research? Is it technique or intuition? In hindsight, I realize that my frustrations were symptomatic of an ongoing debate on the very aesthetics of doing theoretical physics research—things I hadn't learned in the classroom.

In his book *Dreams of a Final Theory*, Steven Weinberg, Nobel laureate and pioneer of the unification of the electromagnetic and weak nuclear force, recalls a talk Dirac gave.

> In 1974 Paul Dirac came to Harvard to speak about his historic work as one of the founders of modern quantum electrodynamics. Toward the end of his talk he addressed himself to our graduate students and advised them to be concerned only with the beauty of their equations, not with what the equations meant. It was not good advice for students, but the search for beauty in physics was a theme that ran throughout Dirac's work and indeed through much of the history of physics.[3]

We can understand why Weinberg partially agreed with Dirac. He was able to use fiber bundle theory, which has very elegant geometric formulas, to discover that the electromagnetic force and the weak force are really just one unified force, the electroweak force. However he did not think that students should strictly follow the advice about not stepping back to ask what the equations actually meant. In Weinberg's Nobel Prize–winning discovery, the equations and calculations he worked on intensely would not have led to his groundbreaking discovery had he not stepped back and asked what the equations meant. Turns out that Weinberg was applying his equations to the wrong physical system, the strong nuclear interaction. In his Nobel lecture, Weinberg says: "At some point in the fall of 1967, I think while driving to my office at MIT, it occurred to me that I had been applying the right [equations] to the wrong problem . . . The weak and electromagnetic interactions could then be described in a unified way in terms of [spontaneous symmetry breaking]."[4]

So what is this mathematical beauty in physics that Dirac is referring to? Many physicists associate beauty in a physical theory with how elegant it is. Look up *elegant* in the dictionary, and you'll find words like *refined, tasteful, graceful,* and *superior.* An elegant equation is *refined,* slimmed down to the bare essentials, simple and concise. An elegant equation is *tastefully* written in the mathematical language of numbers, letters, and symbols. An elegant equation is *superior* in its ability to house within it other equations that can be derived from it. An elegant equation is a beautiful thing.

A good example of elegance in physics is in the equations used to describe planetary motion. Johannes Kepler formulated three laws of impeccable precision that explained the elliptical motion of all the planets around the sun. However, they were missing a key element: gravity. Then Isaac Newton came along. Newton's universal law of gravitation showed how all three of Kepler's equations could come from one single equation. Similarly, Maxwell's equations describing electricity and magnetism, another collection of impressively accurate equations, were gathered up into one single "mother" equation after Einstein showed that space and time could be unified into a four-dimensional space-time. These unifications are pleasing because they simplify the equations.

In subsequent years it was found that *most* particles have an antiparticle, a discovery that had a great effect on one of the grandest theories in physics today—string theory. With its goal of providing a framework for unifying the four known fundamental forces of nature, string theory has been humbly called the "theory of everything." Although string theory is not the only theory to attempt to make this unification, because it stems from elementary particle physics, it has not yet enjoyed experimental success based on the particle symmetries that Dirac unveiled. Its success has, however, highlighted a challenge in the way people think of beauty in physics.

A mathematically beautiful theory is enticing because it provides a playground to explore virtual realities that simulate the real physical world. In string theory, there is not only elegance in its *purpose* of unifying gravity and quantum mechanics but also in *how* it does it. Starting from the equation of a one-dimensional vibrating string, the equations for all the forces—gravity, electromagnetism, and the weak and strong forces—can be derived. A huge feat from simple beginnings. And there is an additional feature, which I find to be very beautiful.[8] It provides an alluring description of the existence of the four forces and matter in terms of the geometry of *extra* dimensions—dimensions beyond our four space-time dimensions.

As beautiful as a theory in physics is, however, it has to measure up to the truth, and the debate often lies in the unexpected predictions a theory makes. String theory, rich with symmetries responsible for its elegance, relies on extra dimensions for mathematical consistency, and it has been argued that string theory predicts the existence of an infinity of worlds—we will discuss this later in the book. Naturally, some physicists do not see this as beauty because it is far beyond our observational capabilities and is outside most physicists' commonsense understanding. Nevertheless, the aesthetic trend, the search for symmetry and mathematical elegance, has continued to impact the discovery process in modern physics.

Out of pure reverence for his genius and contributions to physics, I was one of those young physicists who took Steven Weinberg and Paul Dirac's advice to heart. But my Diracian orientation was exarcerbated by peer pressure and my desire to fit in with the other physicists in my new home. After all, leaving the comforts of six years of close friendships in Providence, I relied on my postdoc colleagues at Imperial for friendship and community. I recall attending a theoretical physics workshop at the Institut Henri Poincaré, hanging around a group of fellow postdocs debating about a string theory topic. Excited, I injected a speculative thought and observed the other postdocs continue on talking with each other as if I didn't exist. Lesson: Show me your chops. Where are your equations? To play in this ballpark, I had to learn the moves—which meant engaging in mathematical gymnastics. This was all too reminiscent of the jazz sessions I had attended where everyone's fixation was on who could best "play" in the chord changes.

Dirac's message was clear: sharpen your technical chops, shut up and calculate, and you'll be successful in the field. So I decided to put my free-flight improvisations and analogical reasoning on hold and try Dirac's method to see what physics fell out of the equations—what reality the elegant mathematics predicted. I wasn't alone; in fact, most of my peers strictly followed this Diracian method of exploration. Behind closed doors, postdoctoral fellows would be hunched over at desks while clocks ticked away. Countless hours were spent calculating, hoping to crack a code that would lead to a breakthrough—or at least to a paper worthy of publication.

I would get my double espresso and march into my office to continue a supergravity calculation. The major goal in cosmology research during my first years as a postdoc was to follow Brandenberger's instruction to find a deep connection with the physics that was operational in the early universe to explain how the large-scale structure in the universe developed. Supergravity was considered to be that theory, in particular an eleven-dimensional version called eleven-dimensional supergravity. Supergravity would have been Dirac's dream because elegantly, in one line, the equation combines the once separate gravitational equations with Weinberg's unified theory of the electroweak interactions. Everyone was convinced that eleven-dimensional supergravity had the right ingredients to search for even more hidden mathematical formulas of beauty and simplicity. I was searching for a hidden pattern in the microworld of supergravity that encoded our macrocosmic structure.

Supergravity is a version of Einstein's theory of general relativity, dressed up with *supersymmetry*—a symmetry that draws a connection between bosons and fermions, matching one boson for every fermion into "superpartners." Bosons are the entities responsible for transmitting forces, such as the photon that transmits the electromagnetic force. On the other hand, fermions are both matter particles, such as electrons and quarks, and the antiparticles of all these subatomic entities. Looking in the mirror is an example of how symmetry works. The image you see closely resembles how you appear to others because of your bilateral symmetry, even though the mirror switches your left side for your right. Think of supersymmetry as a mirror that switches Bosons for fermions without changing the behavior of the physical system.

Conceptually, supergravity was fascinatingly profound, but I was also seduced by the *act* of doing the calculations; the visuals on the paper were beautiful in and of themselves. That moment of awe I had some twenty years back in the Museum of Natural History, staring at the symbols that Einstein had written, had become my reality. By simply raising or lowering the index on a variable, I would be manipulating virtual geometric worlds with the stroke of a pen. The serendipitous moments seemed worth any amount of work. Large numbers of terms would sometimes cancel each other out, simplifying the equations to hand, instantaneously clearing both page and mind. Other times, found patterns in the equations unexpectedly coincided with a known truth of the cosmos, and I'd remember that, for all this Diracian effort, the equations were a reflection of our universe. Though immensely gratifying at times, it was hard work. Postdocs would break for espressos to groan about the futility of their efforts, excitedly share their hallelujah moments, or simply seek caffeine to pump their brains with more calculating power.

Despite my objective to master supergravity and those moments of delight in manipulating equations, months were flying by and my two-year postdoc "hourglass" was running out. I felt that I was getting nowhere closer to figuring out how supergravity and its older cousin, superstring theory, could reveal the secrets of the universe's structure. Although I had nothing to show for my dead-end supergravity calculations, it seemed like everyone else was publishing exquisite papers revealing hidden mathematical structures in the superworld. In spite of my hard work, I was spiraling into a dark well of even more self-doubt. Maybe Robert and Leon had felt sorry for me and had deliberately spared me from entering the mathematical maze, necessary for mastering the technical chops needed to make progress in my research.

Back from a coffee break one afternoon, I got an e-mail from the theory group's administrator, Graziela. The head of the theory group, Dr. Isham, wanted to meet with me. I was scared. "He's figured it out," I thought. "I'm just an imposter." I left my calculations on my desk, stood up slowly, and began heading toward Isham's office, having resigned myself to the possibility of being called a buffoon and being told to excuse myself from his prestigious group.

I first encountered Chris Isham in *A Brief History of Time,* a documentary about Stephen Hawking, in which he, along with Roger Penrose, appeared. Isham, Penrose, and Hawking were colleagues and world-class mathematical physicists. Chris had a mythological aura around him, and he was graced with the rare combination of independent, creative thinking and superhuman mathematical chops. He was able to make key advances in the quantum gravity theories, which try to unify quantum mechanics with gravity—those theories of "everything." In the sixties, he was a young prodigy and the PhD student of Nobel laureate Abdus Salam, best known for the grand unification theory of two of the four forces, not including gravity.

Sadly, like Hawking, Chris suffers from a rare neurological condition that has subjected him to unbearable pain for most of his life. A tall man, you could spot him in the long corridors of Imperial College, beyond the masses of students just out of class. He had an unmistakable hobble to his walk, a sideways lean that looked slightly like a stage comedy act. Admirable, in so many ways, he had a smile and witty humor and wise advice for everyone. An Imperial student shared with me the following story about Chris. Once, on a damp winter day in London, one of those days when the gloom makes you question if it's really necessary to drag yourself out of bed, Isham decided to shock his students into wakefulness. Never wanting for new ideas, he announced briefly that he would conduct the written part of his lecture backward. "It's a convention! Why read from left to right if you can read from right to left?" he asked, with a big grin. The surprised class quickly woke up and didn't waste time taking notes. That day's lecture was on fiber bundles, one of Isham's favorite topics at the time. He made it look easy on the blackboard. Math was in every cell of his body, so any which way it flowed—back to front, upside down, whatever—didn't much matter to him.

I walked into his spacious office and saw Dr. Isham relaxing on a reclining armchair with his feet up. His arms were trembling slightly. Notes on topos theory—incredibly complicated algebraic-type manipulations of rules on topological spaces—decorated the board behind him, so grand, they couldn't possibly fit on A4 paper. He was smiling warmly. He wasted no time and got straight to the point. "Why are you here?" he asked. I responded, with some nervousness in my voice, "I want to be a good physicist." Chris then surprised me. "Then stop reading those physics books. You need to develop your unconscious mind; that's the wellspring of a great theoretical physicist." As if his scientific repertoire weren't impressive enough, what I didn't know at the time was that he was both seriously spiritual and philosophical. He calmly and earnestly told me that he had trained his mind to do tedious calculations while he was dreaming. He followed that remarkable revelation with another question: "What are your hobbies?" Dumbstruck by his feats during slumber (I just *slept* at night), I distractedly replied, "I play jazz at night." There was a pause. "You should play more music. I sing. I find that music is an ideal activity to engage the unconscious." Another pause. "You see these books here?" He pointed to the complete volumes of Carl Jung's writings, the founder of analytical psychology. "I have fifteen years of training in Jungian psychoanalysis. Read volume nine, part two, *Aion: Researches into the Phenomenology of Self.* There is a mystical side to doing physics. Do you know that Pauli and Jung worked together?"

I was amazed at this possibility. Wolfgang Pauli was nominated by Einstein for the Nobel Prize for his work as one of the architects of quantum mechanics. Pauli, who was known for quickly catching mathematical mistakes and forcefully objecting to sloppy thinking, was famous for his statement: "Your theory is not even wrong." He was a true Diracian. He also predicted the existence of a very elusive particle, the neutrino, and was known to be a technical force to be reckoned with. I found it hard to believe that Pauli would be involved in psychobabble, but my perspective was about to take a turn. Chris pointed to another book, *From Atom to Archetype*, a collection of letters exchanged between Pauli and Jung for over two decades. "You can see that Pauli came up with the Pauli spin matrices from a symbol that appeared in his dreams. Do you want to borrow the book?" I didn't just borrow it; I treated it like I'd just uncovered a precious font of insights. It was a new groove, which called for new digs. Instead of numerical grumblings in the theory group café over black coffee, I immersed myself in reading the collection of letters over pints in a pub on Portobello Road.

Privileged to be under his wings, I met with Chris weekly to discuss foundational problems in theoretical physics, and I followed his advice like a disciple. I joined a jazz trio and had two regular residencies or gigs at venues in Notting Hill. I read Jung and told Chris the dreams I had. After months of this, my new habits paid off. I'd been stuck on a project. I was trying to connect string theory to cosmological inflation, the idea that the newly born universe went through a rapid, or inflationary, phase of expansion just after its birth. One night, in the middle of a sax solo on Coltrane's song "Mr. PC," an image appeared in my mind that I *knew* had to do with the resolution of my project. The next day, I woke up with this newfound insight and ran to my desk to jot down some equations that related to a branch of mathematics called noncommutative geometry, which proposes that the speed of light in the early universe could vary. With cosmologist João Magueijo, I ended up publishing a paper on this new connection, which has been cited over one hundred times. It did not get me any closer to my supergravity-cosmic structure dream, but I started to see that Chris's system was working for me.

Between the encouragement to play music from both Branden-berger *and* Isham, and the respect for the mathematical excellence and elegance they were both capable of, my Diracian method began to transform. "Playing" with equations was now a multifaceted task. Although it is essential in jazz to practice your scales and technique, it is also important to get out and play with others. Improvisation is an in-the-moment act, and the only way to prepare for it is to get out and play. I started to realize that I was missing this in my research. I was too literal in my approach to the Diracian method. By taking my research questions and challenges into the jazz sessions, I found myself like a child playing in a sandbox, without a care of being wrong or silly. After so many years of loving both jazz and physics but keeping them on separate playing fields, suddenly music was developing my uncon-scious mathematical muscles. What I had first seen as psychobabble had become an avenue of productivity. Pauli's conversations with Jung were, after all, what led Pauli to discover a new property of matter and a new law of nature. Since my college days, ideas connecting music and cosmology had been stirring in the back of my brain, and now I was digging them out of my unconscious, facing them, and thinking they were not quite so outlandish as they had seemed before.

Playing music with others became part of my scientific practice. It was a way of doing research, and it was also a lot of fun. During the overnight jazz sessions during the summers in New York City, I'd bring along my physics calculations. During breaks, when the jazz cats would talk shop, I'd bring up my physics research and make connections to jazz. I still engaged in the Diracian method; I just changed my environ-ment. I started to become more playful with the new formalisms I was exploring in my research in the sessions. I would look at the trade-off between the trumpet and the sax solos as two superpartners undergo-ing a supersymmetric inversion. I will never forget telling a now well-known pianist about how his playing was geometric. A fellow tenor player jumped in and said, "I don't know what you talking about. His stuff just sounds good!"

A key geometric concept is isometry—preserving the distance between a set of points. A simple case of what I meant by "geometric" in
the pianist's solo is analogous to taking a square on a two-dimensional
surface and sliding the square around. The shifted square is isometric
to the original square because the distance between the four edges is
preserved. There are certain surfaces that are so warped that moving the
square around would not preserve the distances between its four sides.
Oftentimes, a jazz soloist will repeat a melodic and rhythmic pattern,
or riff, and practice it in all twelve keys: this shifting of the patterns
into different keys preserves the tonal distance relationships between
the notes in the melodic pattern.

Years later, I discovered I was not that crazy to make associations between geometric reasoning and symmetry and music back in my days
at Smalls. And it's worth discussing a remarkable similarity between
how symmetry and the use of geometric reasoning functions in music
and in improvisation. Let's turn our attention to Stevie Wonder's "You
Are the Sunshine of My Life" and Debussy's "Voiles." Those songs share
something powerful in common (aside from being beautiful pieces of
music). They both use a symmetric scale, so called because it looks symmetrical when you draw it. Consider the twelve notes written in a circle: if we start with the note C and incrementally go up sequences of
half steps twelve times we end up with C. This is a geometrical representation (a circle with discrete points) of the twelve musical notes.

We can draw lines that connect various notes to one another. The
same way an equilateral triangle is the most symmetrical three-sided
object in two dimensions, the hexagon is the most symmetrical six-
sided object in two dimensions. The notes that land in the hexagon in
Figure 4.1a—C, D, E, F-sharp, A-flat, and B-flat—comprise a symmetric scale called the whole tone scale. What does it sound like? You can
hear it at the beginning of "You Are the Sunshine of My Life" when the
electric keyboard plays some airy ascending notes before the vocals begin. In Debussy's "Voiles," this whole tone scale is descending.

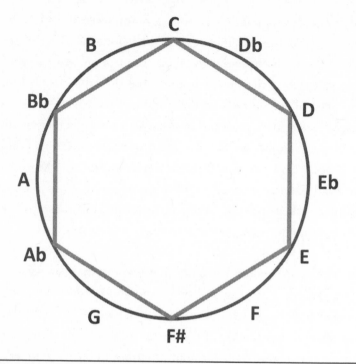

FIGURE 4.1A. The sixfold symmetry of the whole tone scale.

We can make other symmetric scales, such as a perfect square, or diminished scales. Pat Martino, considered one of the greatest living jazz guitarists, developed a deadly brain aneurysm in 1980. Surgery saved his life, but as a consequence, he suffered amnesia and had to completely relearn the guitar. Coincidentally, he developed a new system based on the symmetry of just two "parental forms" that are innate to the geometric structure of the guitar's fret board. These parental forms or chords are nothing more than the symmetric augmented triad and diminished seventh. Martino gave many presentations demonstrating the power of the parental chords to each generate a wide array of chord offspring.

Likewise, as we will soon discuss in some detail, John Coltrane also used symmetric scales as the basis for his album *Giant Steps*. When he told David Amran he was inspired by Einstein to find a simple idea for his music, that idea, drawn from reading Einstein's books on relativity, was symmetry. When I studied Coltrane's mandala, I realized it was screaming symmetry. Consider the space-time symmetry between inertial observers that led Einstein to special relativity and the four-dimensional version of electrodynamics. He further used the symmetry of curved space-time to show the equivalence between an accelerated observer and a static observer. From the unifying and simplifying idea of symmetry, Einstein was able to express disparate and complicated physical ideas in one fell swoop.

I had the great fortune to speak with Martino, a kind man who spoke with unfathomable logic. He told me that he wanted to find a system that expedited his improvisational ideas and technique. I couldn't help but smile. "A lot of physicists also try to find the most efficient theory that can account for as many phenomena in one shot," I said. Martino and Coltrane coincidentally were both trained by the late Philadelphia jazz guitarist Dennis Sandole, who introduced them to the concept of symmetric scales, or what he called "equal divisions of the octave and third relationships."[5] But what Trane and Martino did not realize was the deep connection between symmetric scales, their symmetry breaking, and physics. In the breaking of symmetric scales we can hear the physics of symmetry breaking.

Chords usually refer to a "tonal center." For example, a C-major chord, comprised of notes C, E, and G, has C as its tonal center. If you play that chord, C will be the most prominent note you hear, while G and E will serve to add harmony and embellish the C note. But a symmetric chord has more than one tonal center, which makes it sound ambiguous: the notes are evenly spread out and refer to several tonalities. For example, the C whole tone chord has all five notes—C, D, E, F-sharp, A-flat, and B-flat—as its tonal center. Because of their uncertain sound, symmetric scales have a special role in music. For hundreds of years, composers like Ravel and Bach have used them to create tension or a sense of ambivalence, usually when a piece is about to make a harmonic change. Another ubiquitous symmetric scale is the diminished scale, which is widely used in jazz music and in the compositions

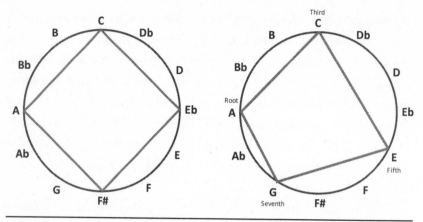

FIGURE 4.1B. A demonstration of symmetry breaking of the symmetric diminished scale *(left)* to a major scale *(right)*.

of Gershwin and Cole Porter to create movement between two chords with tonal centers.

Figure 4.1b shows the symmetric diminished chord to the left and a major chord to the right. If you look closely, you will see that the diminished scale can become the major scale by only changing two out of the four notes. In a symmetric scale, there is no hierarchy among the various tonal centers. By contrast, an asymmetric major scale breaks that symmetry and establishes a hierarchy, resulting in a tonal center—the major chord—that becomes more important than the other notes in the scale. We will return to this important musical analogy when we explore how asymmetry in the early universe similarly established hierarchies in the earliest cosmic structures.

All those years I spent working toward my PhD, then as a postdoc, and now as a professor doing independent research, other jazz musicians were in conservatory, committing thousands of hours to mastering music theory and instruments. When among physicists, I usually kept my curiosity about the relationship between music and physics to myself. But I could not shake my compulsion. After all, Dirac's way said that mathematics, and in particular the use of symmetry, led to progress in fundamental physics, and I had experienced for myself the geometry and symmetry of music in those improv jazz sessions at Smalls. Plus, the flashes of insight about physics that would hit me sometimes when I soloed on my tenor I knew were for real. Years later, I was comforted to learn that the very genesis of what we know as physics was sparked by a group of people who sought elegance, and they found it in the union between music and mathematics—the Pythagoreans. It seems I'd been bit by the Pythagorean bug.

5

PYTHAGOREAN DREAM

During all the years of formal training as a scientist, I found myself trying to reconcile my passion for music with physics. I started to see not only how the act of doing physics research could benefit from musical analogies but how our physical world actually had a musical character. Aside from the few mentors, such as Chris Isham and Robert Brandenberger, who had encouraged me to blend the two, I felt pressure to keep these two worlds separate. Physics to some is about absolute truths encoded in rigid mathematics, and music is a language of emotion. Perhaps this tension would not have been a big deal had I known that in the early days of science, music and astronomy were inseparable. To the modern musician and scientist, this may seem preposterous, but to early people, who lacked the scientific tools we now have, music became an analogy for the ordering and structure of the cosmos.

Like modern folk, ancient people also wondered where they came from and what their place in the universe was. The awe that people, ancient and modern, feel when faced with questions of birth and death and with the challenges of the natural environment, its bounty and adversity, inspired people to personify nature and to worship and appease those natural forces. Creation myths are largely a by-product of these factors. The shift from creation myths to deductive, scientific-like reasoning most likely began with the Pythagoreans over twenty-five hundred years ago. They sought a combination of mathematical and mystical agency to understand the motions of celestial bodies and humanity's relation to them.[1] Although it is arguable that the notion that the universe is mathematical originated in Babylonia or Egypt, credit is given to Pythagoras for positing that "the universe comprises of harmony arising out of number." If only I had been aware of this history in my days of string theory research, it would have supported what my teachers of the time were encouraging me to do, combine science and music. But, because intense interdisciplinary study was an uncommon practice, I doubted its validity. In retrospect, the Five Percenters were on to more than I realized at the time.

Pythagoras is best known now for his famous Pythagorean theorem, which lets you determine the hypotenuse (h) of any triangle with a 90 degree angle if you are given the lengths of the two other sides (a) and (b).

Although it is debatable whether Pythagoras actually deserves credit for the formula: $a^2 + b^2 = h^2$, this is not his only claim to fame. Many would be surprised to learn that he is responsible for none other than our Western music scale. He used mathematical deduction to comprehend the physical world and set the stage for the enormous breakthroughs that occurred in astronomy and physics millennia later.

Legend has it that Pythagoras left his home in search of divine knowledge. From the island of Samos in the eastern Aegean Sea he traveled to Egypt and Babylonia and returned, some two decades later with the conviction that all of creation existed in a perfect harmony of numbers. The orbits of the planets, he proposed, played musical notes, which he called tones, whose pitch was dictated both by the speed of the planet and its distance from the sun. At the time, there were five known planets, which Pythagoras postulated together played a beautiful harmony. The stars, in *their* positions and movements, were also believed to contribute to the cosmic song, and legend has it that Pythagoras could hear this "music of the spheres." To the Pythagoreans, truth itself was in numbers and their relationship to each other. Mathematics was the way to unravel the universe's secrets, and the harmony of the cosmos was, simply, a manifestation of the relationships between numbers.

As the story goes, Pythagoras's eureka moment was at a blacksmith's shop. With the repeated hitting of their hammers, metal against metal, Pythagoras honed in on what is now known as a consonant tone. His sensitive ear and mathematical mind were able to pick out sound vibrations, or notes, that were pleasing to the ear—sound vibrations that resonated with, or vibrated in sync with, the actual physical structure of the ear. Upon inquiry, the blacksmiths revealed that the weight of the hammers differed from each other by ratios of one-half.

With his heart immersed in the possible harmonic nature of the cosmos, he got to work on applying these ratios to his observations and postulates. He set up an experiment with suspended strings that, when plucked, generated tones. The ratios between the weights of the hammers, he realized, were equivalent to ratios among the string lengths. Despite being different objects of different media, the mathematics and harmonics were the same. He was on to something big. When he plucked a string half the length of the original, he would have a similar tone but of a higher frequency—this is called an octave. Repeating this process of vibrating strings with lengths that were successively divided in halves led Pythagoras to the discovery of our Western musical scale. Halving the length of a string was equivalent to doubling its frequency. The ratio of 1:2 would be equivalent, for example, to a 220 hertz A note frequency as compared to a 440 hertz A note frequency. In Western music, this doubling of frequency is equivalent to rising by an octave, from one A to another, or moving up eight keys on a classical piano. We perceive an octave as identical but at a higher pitch from the original note.

Pythagoras further plucked the string at one-third of its original length and heard it vibrate at a perfect fifth, which in the key of C is the G note. That's the second note in Elvis Presley's melody "I Can't Help Falling in Love with You": "Wise (C) men (G) say (E)." When the string is divided into one-quarter of its original length, we get an F note in the key of C. This pattern continues for all whole integers from one

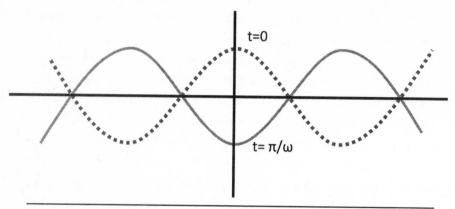

FIGURE 5.1. String vibration.

to five. The fact that the consonant tones were generated from integer relations of the length of the string vindicated Pythagoras's conviction "that all is number" and that the music of the spheres existed.

Although Pythagoras was able to show that the ratios of small numbers in the length of a string would generate consonant tones, *why* this was the case remained a mystery, and we will dedicate an entire chapter to this topic. His discoveries were not only the bedrock of much breathtaking music to come, from the likes of Bach, Mozart, and the Beatles, but they were also an important contribution to pure mathematics and astrophysics. The Pythagorean's fundamental belief that all is number has become, almost three millennia later, the mantra of modern theoretical physics. Dirac was, I found out, a Pythagorean.

Ancient Greek philosophers and astronomers believed that Earth was at the center of the universe—that it was a geocentric universe. After all, the force of gravity had not been discovered, and it did appear that everything fell toward Earth. Surrounding Earth were perfect spheres governing the motion of the spherical celestial bodies; hence the music of the *spheres*. Equivalent to the divine, the circular form was faithfully recognized as the essence of the structure and dynamics of the universe, and everything was done to maintain that this was so.

Aristotle's (ca. 350 BCE) revered model of cosmic perfection is appreciated even today for its beauty, if not for its accuracy. The planets, stars, and moon were embedded in crystal spheres rotating around Earth; spheres were made, he said, of a fifth element called ether. Aristotle was a student of Plato, who was himself a Pythagorean. Plato had extended the numerical basis of the music of the spheres to geometric shapes, celebrating the platonic solids, which carry his name. The five platonic solids, next to the sphere, were the most special geometric shapes in their symmetry, regularity, and precision. According to Pythagorean-platonic philosophy, these perfect geometric forms were understood to be autonomous from the human sphere of existence, as was musical perfection.[2] Each of these convex solids is made up of one type of regular polygon, with the same number of polygons meeting at every vertex. The cube is perhaps the simplest and most well-known platonic solid. It is made up of six squares, three meeting at every vertex. The others are the tetrahedron (four triangles), the octahedron (eight triangles), the icosahedron (twenty triangles), and the dodecahedron (twelve pentagons).

Plato associated these forms with the four elements: earth (cube), fire (tetrahedron), air (octahedron), and water (icosahedron). As no element was associated with the dodecahedron, it is debatable whether Plato actually discovered all five. The fascination of ancient philosophers with the beauty of the cosmic realm, and the attempt to find mathematical precision to match it, sparked the birth of what we know as modern science.

Four hundred years after Aristotle, more precise astronomical observations brought the old cosmological model under scrutiny, and the ancient crystal spheres began to shatter. Ptolemy (ca. 100) created his famous Ptolemaic model to explain the apparent retrograde motion of planets as they moved through the night sky. We now know that the planets, as seen from Earth, occasionally exhibit a slowing down and backing up motion as they traverse the sky due to their motion around the sun relative to Earth's. But in the geocentric, perfectly circular models that existed, it was nearly impossible to justify irregular

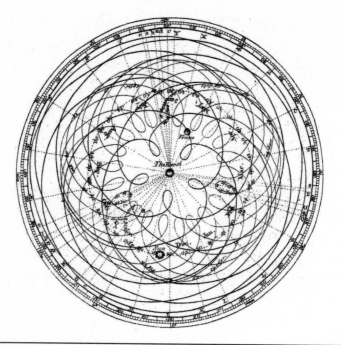

FIGURE 5.2. Image of Ptolemaic model. *Wikipedia.*

motion such as this. Ptolemy, a creative genius, did his best with his Ptolemaic model. Cosmic divinity, and its circles of perfection, were not yet considered adjustable, so he had to introduce circular cycles *within* cycles, or epicycles, to explain not only the retrograde motion but also the slightly elliptical motions that some of the planets seemed to exhibit. It was a terribly complex and unintuitive model necessary to uphold ancient beliefs.

It took about fifteen hundred years for the belief in divine geometry to begin to give way to confidence in human perception. It wasn't easy for a mere mortal to stand up to the divine, and some would suffer the consequences. Enter the supernova of 1054. It was an incredibly bright exploding star, responsible for the spectacular Crab Nebula observable in the sky today. Recorded by both Chinese and Arab astronomers, it suddenly appeared in the night sky among the fixed stars. Something new, something changeable, in God's perfect universe? A cacophony in

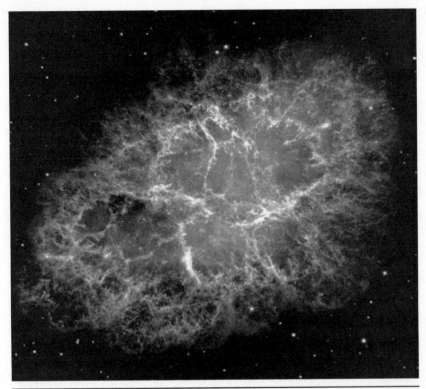

FIGURE 5.3. The Crab Nebula, a remnant from an exploding star (supernova). NASA, ESA, J. Hester, A. Loll (ASU).

the music of the spheres! The supernova was a clear anomaly, yet it still took hundreds of years for someone to overthrow the ancient standards. Nicholas Copernicus, a Polish mathematician and astronomer born in the late 1400s, threw the stone that shattered the crystal spheres, epicycles, and God (for some), once and for all.

Copernicus displaced Earth with the sun, and the universe became heliocentric, with all the planets orbiting the sun. Retrograde motion of the planets was a result of our perspective of the orbiting planets from Earth, and the inconstant motion of the sun across the sky was due to the annual revolution of Earth around the sun. The daily motion of the stars across the sky, he said, was due to Earth's rotation on its axis. What an insight! Not only did he accurately describe all of these points, but he also identified that the stars were much farther away from either the

planets or the sun, explaining the difference in their motions. Heliocentrism has thus been appropriately exalted with the title the Copernican Revolution. Perhaps luckily, perhaps out of stress, Copernicus died
the same year he published his sun-centered theories of the universe.
He did not live to see the consequences of his demotion of God to an
onlooker in a more human and cerebral cosmos.

Galileo Galilei (ca. 1600), the father of observational astronomy,
unveiled some of astronomy's greatest eye-candy. Sunspots, the moon's
craters, and the phases of Venus were all carefully observed and identified by him. He also delighted in the Milky Way, that magical band of
dense stars and interstellar clouds that divides our night skies as we gaze
toward the center of our galaxy. Perhaps his most significant discovery,
however, was of the four largest moons of Jupiter, now known as the
Galileans. Meticulously recording, night after night, hour by hour, the
way the planets appeared to oscillate back and forth in front of, and
behind Jupiter, he was able to prove that these moons revolved around
Jupiter. This observation alone was enough to disprove geocentrism and
provide strong support for Copernican heliocentrism. Unlike Copernicus, he lived to support his arguments, was duly accused for upsetting
the divine order, and was placed under house arrest for the remainder
of his life. Though punished for messing with God, Galileo and his
predecessor, Copernicus, had carved a path that would remain open
for future theorists wanting to find physical, not just divine, reasoning
behind the beauty they saw.

Paralleling Galileo in the feats of observational astronomy, but more
conservative in his choice of beliefs, was Tycho Brahe (late 1500s), a
Dutch astronomer and nobleman. He devoted much of his career to in-
venting instruments and took progressively more precise measurements
of the positions of the heavenly bodies. He undoubtedly had the most
accurate data on the motions of the heavenly bodies to date. Though
he respected some of Copernicus's geometrical arguments, he was not
ready to give up geocentrism, and stayed strictly Ptolemaic. In Decem-
ber of 1599, he informally invited a young assistant, Johannes Kepler,
to help him catalogue all of his data. But one year later, he suffered an

unprecedented death, leaving his vast collection of detailed records to his family. Rather reluctantly, they eventually conceded to pass Brahe's precious work to his recent assistant, the young Kepler.

Johannes Kepler (ca. 1600) was, arguably, the first astrophysicist, seeking a physical, as opposed to a divine, cause for the motion of the planets. He had a tragic life. His mother was almost burned at the stake for witchcraft, he saw the last of his mercenary father at the age of five, and he lost three children and his wife to plague and illness. He himself had smallpox as a child, which left him with impaired vision and crippled hands. With all of Earth's tragedies, it was from its surface that Kepler, as a young boy, observed great and inspiring cosmic phenomena. The suffering and grief only worked to accelerate his yearning to embrace the heavens. The first significant event was the great comet of 1577. It passed close to Earth and was observed all over Europe—the same one Brahe observed, which led him to correctly deduce that comets were not an atmospheric effect, but an occurrence beyond the realm of our planet. Later, Kepler observed a lunar eclipse. Despite poor eyesight, he was struck by the moon's reddish appearance and permanently turned his heart, and what was left of his vision, to the skies. Brahe's eyes, it turned out, were the perfect complement to those of his less fortunate assistant's.

Kepler, a brilliant mathematician, a passionate astronomer, a creative experimentalist, and an individual bold enough to deny geocentrism and spherical perfection, followed a strict academic career and did justice to Brahe's life work.

One day while giving an astronomy lecture, Kepler came to a stunning flash of insight. When discussing the motions of the planets, Kepler realized that the distances between the planets were not accidental and reflected the Pythagorean view of a divine rational order. For instance, Mars, half as far as Jupiter, would orbit an octave higher. He was certain the planets moved to the music of the spheres, but like a modern scientist, he kept asking questions. Why, he wanted to know, were there six planets? Why not twenty-five or three? Uranus and Neptune had not yet been discovered. After days of frustrated laboring, Kepler had an astounding revelation that would stay with him for the rest of his life—the universe must obey some deep geometric harmony in its mathematical ratios. The five platonic solids, he realized, were the very reason *why* there were six planets, and they could also dictate the separation and motion of the six planets.

Think of a Russian doll made out of the platonic solids. Each platonic solid can be placed within a spherical shell such that every vertex of the solid touches the sphere's inside surface. Similarly, a sphere can be placed within each solid, touching the inside faces of the solid. In this way, Kepler was able to construct a model that posited six imaginary spherical shells, which he associated with the location of the six planets. Risking the fate of Galileo, he placed the sun at the center, followed by Mercury, Venus, Earth, Mars, Jupiter, and Saturn. Their movements were limited to the spherical orbs separated by the platonic solids. The results of his findings were published in the book *The Mysterium Cosmographicum* (The Cosmic Mystery) in 1596. It was a remarkable feat that using symmetry and geometry, Kepler was able to solve two fundamental problems that had loomed over the heads of philosophers and astronomers for two thousand years. At twenty-six years of age, Kepler had figured out why there were only six planets *and* positioned their orbits—nearly.

All was not quite right. Graced with incredible intuition and the mathematical drive to be constantly refining his equations and models, Kepler was left with a profound question—what, he wanted to know, physically *caused* the planets to go around the sun? This was a new category of inquiry. A religious man, Kepler initially sought an explanation in the Holy Trinity. God, the sun, was located at the center; the Son was the fixed stars; and the Holy Spirit emanated the power, or force, necessary to create the motion of all the celestial bodies. But this was not enough of an explanation for him. Kepler wrote that "either the souls which move the planets are less active the farther from the Sun, or there is only one moving soul in the Sun, which drives planets the more vigorously the closer the planet is."[3] He further reasoned that the

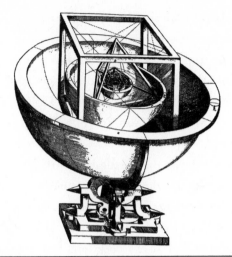

FIGURE 5.4. Johanne Kepler's *Platonic solid* model of the *solar system* from *Mysterium Cosmographicum* (1596). *Wikipedia.*

power "diminished in inverse proportion to distance as does the force of light." This was the first time in history that physical reasoning was applied to astronomy. What Kepler was slowly unraveling was not trivial. Underlying these nagging unknowns were both the physics of gravity and of light. Similar in action, he was correct that they both dissipate the farther one is from the source. Kepler also had the correct intuition that the sun's "soul" was a force driving the planetary motions—indeed, a force that would later be called gravity. But still, he needed more information, which he would find was already hidden within his grasp. Scrawled in a busy and attentive hand were Brahe's records on the motion of the planet Mars.

Not one to overlook detail, Kepler was quick to realize that Brahe's data on the orbit of Mars was not fitting his model as precisely as he'd hoped. Mathematical rigor would certainly iron out the details. It might take days, he thought, but that didn't deter him. It took almost *eight years*. The problem was that Mars's orbit around the sun deviated so much from a circle that it was impossible to reconcile it with a model of perfect spheres. Unlike other astronomers before him, Kepler finally compromised his idealized model of platonic solids in the spirit of developing what is now considered the modern scientific method of hypothesizing and testing in order to acquire new knowledge. It was known at the time that Earth emanated a magnetic force, which ultimately led Kepler, by analogy, to replace the Holy Spirit with a magnetic force emanating from the sun. In 1605, Kepler wrote:

> I am much occupied with the investigation of the physical causes. My aim in this is to show that the celestial machine is to be likened not to a divine organism but rather to a clockwork . . . insofar as nearly all the manifold movements are carried out by means of a single, quite simple magnetic force, as in the case of a clockwork[.] all motions [are caused] by a simple weight. Moreover I show how this physical conception is to be presented through calculation and geometry.[4]

Though it was true the sun had a magnetic force, which was responsible for the sunspots observed by Galileo, it was not the gravitational force Kepler was looking for.

Continuing his investigations, he decided to revisit his Pythagorean ideals and his faith in the harmony of the spheres. He had been well exposed to the beauty that came from such ideals. After all, Pythagorean mathematics had defined Western music, and famous Renaissance composers such as Monteverdi and Byrd were contemporaries of Kepler. He reasoned that, if the planets moved in a perfect circular orbit, the pitch would remain the same throughout the orbit. But since in a case like that of Mars, that orbit was apparently oblong, Kepler argued that when the planet is closer to the sun, it moves faster and the pitch goes higher. These pitch changes contributed in a new way to the Pythagorean cosmic harmony. While the Pythagoreans believed that ratios produced consonant notes, such as the octave (2:1) or the perfect fifth (3:2)—the fifth note in "do re mi fa *so* la ti do"—the key to Kepler's version of celestial harmony was to take the ratio between the largest and smallest orbital velocities of the planets.

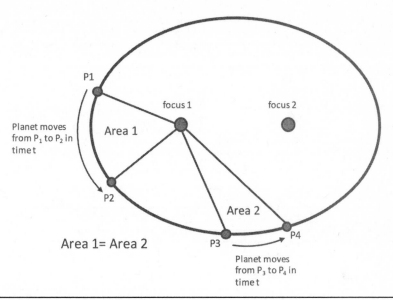

FIGURE 5.5. Kepler's second law.

It is fascinating that geometric and musical reasoning led Kepler to what is today known as Kepler's three laws of planetary motion—nontrivial equations dictating the precise motion of the elliptical orbits of the planets. By 1605 he had determined that planets move in elliptical orbits, not some other oblong shape, and that a line joining them to the sun would sweep out equal areas of space in equal periods of time (see Figure 5.5). It was not, however, until 1620, *fifteen* years later, that he published all three of his planetary laws to include not just Mars but *all* of the planets. The last law drew a precise mathematical relation between the planet's orbital period and the size of its orbit.

It was a long road to success, but Kepler was finally able to realize Pythagoras's legendary celestial music and even write down the notes of the score to share with the world. I recently had the pleasure of listening to an album of Kepler's celestial music called *The Harmony of the World: A Realization for the Ear of Johannes Kepler's Astronomical Data from Harmonices Mundi 1619,* by Yale composers Willie Ruff and John Rodgers.[5] The recording provides an auditory representation of

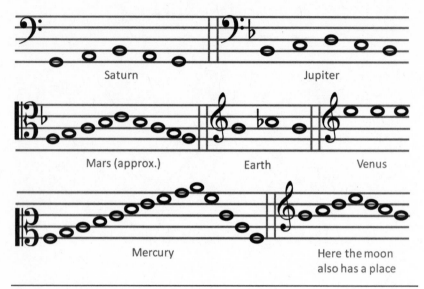

FIGURE 5.6. The musical notes that Kepler calculated for each planet's elliptical orbit. For each planet, the lowest note corresponds to the largest distance from the sun (the lowest orbital velocity), and the highest note corresponds to the shortest distance from the sun (the highest orbital velocity).

the haunting and mesmerizing harmonies the planets make as they go around the sun in elliptical orbits. One surprising feature is how the harmonies blend to create a unified rhythm corresponding to the periodic orbits of the planets. The planet Saturn, for example, played a major third (a pitch ratio of 5:4), Jupiter a minor third (6:5), and Mars played the fifth (3:2). In Kepler's vision, all the planets sang together in a celestial harmony at the pleasure of the divine.[6]

Ultimately, it took a different sort of genius to nail down the correct physics behind Kepler's laws. Isaac Newton (late 1600s)—hands down, one of the most influential mathematical physicists of all time—was able to figure out that a new force, gravity, was needed to attract the planets to the sun and hold them in their elliptical orbits. In his *Mathematical Principles of Natural Philosophy*, published in 1687, Newton described the motion of both celestial bodies *and* objects on Earth due to gravity. What's more, he was able to derive Kepler's three laws of motion from his own universal law of gravitation, the very laws Johannes Kepler, bit by the Pythagorean bug, had arrived at through the analysis of geometry and harmony in the cosmos.

Jumping from one domain to another, Kepler clearly recognized the benefit in analogous thought: from spherical divinity, to the mathematics of harmonic and geometric proportion, to Earth's magnetic force and the unfamiliar wobbles of astronomical bodies, Kepler remains an inspiration to any researcher. Particularly to theoretical physicists, his dedication, creativity, and mathematical rigor demonstrate the ultimate avenue to discovery.

Today, well beyond the era of the Pythagorean dream of harmony in the spheres, the idea that planets play notes may come up against speculation and be regarded as irrelevant. We can imagine, though, the ecstatic fun Pythagoras would have with current knowledge of the existence of so many planets beyond our own solar system. As of October 2012, the Kepler mission had discovered over eight hundred solar planets outside our own system—on the order of a trillion in our own galaxy now—not to mention all the new objects in the universe discovered since Newton's lifetime: galaxies, clusters of galaxies, and elusive dark matter. There is, of course, also the subatomic realm of quarks and neutrinos, and all the symmetry it incorporates in unifying different particles and forces. What would Pythagoras say at the possibility of *strings* underlying all of nature? The universe today would be a dream of infinite geometric and harmonic possibilities for Pythagoras. But what about us?

Modern-day physicists are very aware that their beautiful mathematical models fall short of describing what they see. A physicist's job is never done. There is always a discrepancy, a disagreement, an unjustified element or initial condition, or an anomaly and certainly always more questions. While some physicists see this as an opportunity to discover a deeper mathematical truth, others believe that we have reached a limit in our knowledge. In facing these limits, Pythagorean ideals could prove useful. I have wondered, given what we know today, if the intuition of the musical nature of the universe could be applied to today's puzzles to take us to a new vista in modern physics. Could the cosmos, in fact, be a vast harmonic realization of vibrations? And what is the role of dissonance in our universe? Certainly, we will run up against a limit to any analogy drawn, but in the pursuit of knowledge, the power of analogy lies in its limitations. Knowing that analogies are bounded, by definition, carves new paths for discovery, so long as one is bold and creative in facing the questions they leave us with. Elegance and beauty lie not only in the form of equations but in the method of human discovery itself. Leon Cooper, Robert Brandenberger, and Chris Isham were immensely influential teachers throughout my education because of their methods of discovery.

Students of ancient philosophers pondered the music of the spheres, perfection in geometric form, dynamism, and the organic human being versus the mathematics of the universe. Present-day students are trained in the precise calculations bred from these ancient philosophers—the elliptical orbits of Kepler, Newton's gravitational laws, and Einstein's more complex space-time calculations. What students of the future will be studying is a complete unknown. Education, technology, and global interconnectedness are all developing at enormous rates. For the student to keep up, for the researcher to discover new truths, and for the professor to lend guidance and insight, it may take a combination of ideas from ancient and modern-day philosophy, as well as creativity and improvisation, with the bold willingness to make mistakes.

For me there was no going back.

6

ENO, THE SOUND COSMOLOGIST

It's intuitive to think that anything complex has to be made by something more complex, but evolution theory says that complexity arises out of simplicity. That's a bottom up picture. I like that idea as a compositional idea, that you can set in place certain conditions and let them grow. It makes composing more like gardening than architecture.

—Brian Eno

Everyone had his or her favorite drink in hand. There were bubbles and deep reds, and the sound of ice clinking in cocktail glasses underlay the hum of contented chatter. Gracing the room were slender women with long hair and men dressed in black suits, with glints of gold necklaces and cuff links. But it was no Gatsby affair. It was the annual Imperial College quantum gravity cocktail hour.

The host was dressed *down* in black from head to toe—black turtleneck, jeans, and trench coat. On my first day as a postdoctoral student at Imperial College, I had spotted him at the end of a long hallway in the theoretical physics wing of Blackett Lab. With jet-black wild hair, beard, and glasses, he definitely stood out. I said, "Hi," as he walked by, curious who he was, and with his "How's it going?" response, I had him pegged. "You from New York?" I asked. He was.

My new friend was Lee Smolin, one of the fathers of a theory known as loop quantum gravity, and he was in town considering a permanent job at Imperial. Along with string theory, loop quantum gravity is one of the most compelling approaches to unifying Einstein's general relativity with quantum mechanics. As opposed to string theory, which says that the stuff in our universe is made up of fundamental vibrating strings, loop quantum gravity focuses on *space* itself as a woven network of loops of the same size as the strings in string theory. Lee had just finished his third book, *Three Roads to Quantum Gravity*, and was on a mad rush to mail out the manuscript to his editor. I accompanied him through the drizzle to the post office and for a celebratory espresso—the first coffee of hundreds we'd share in the future.

Lee had offered up his West Kensington flat for the quantum gravity drinks that evening to give the usual annual host, Faye Dowker, a break. Faye enjoyed being the guest lecturer that evening. Slim, bespectacled, and brilliant, she was also a quantum gravity pioneer. While Professor Dowker was a postdoc she studied under Steven Hawking, working on wormholes and quantum cosmology, but her specialty transformed into causal set theory. After a couple of hours, the contented chatter gave way to Faye as she presented her usual crystal-clear exposition of causal sets as an alternate to strings and loops. Like loop quantum gravity, causal sets are less about the stuff in the universe and more about the structure of space-time itself. But instead of being weaved out of loops, space-time is described by a discrete structure that is organized in a causal way. The causal-set approach envisions the structure of space analogous to sand on a beachhead. If we view the beachhead from afar, we see a uniform distribution of sand. But as we zoom in, we can discern the individual sand grains. In a causal set, space-time, like a beach made up of sand, is composed of granular "atoms" of space-time.

Scattered into the quantum gravity mixer were those working primarily on string theory, like the American theorist, Kellogg Stelle, who was a pioneer of p-branes, as well as one of my postdoc advisors. In mathematics, a membrane is a two-dimensional extended object—that is, it takes up space. A p-brane is a similar object in higher dimensions.

The strings of string theory can collectively end on p-branes. And coming at quantum gravity from a yet another route, there was Chris Isham, the philosophical topos theory man who played with mathematical entities that only "partially exist." Postdocs studying all avenues of quantum gravity filled in the gaps between the big brains in the room. It wasn't exactly a gathering of humble intellect. It was scenes like that, that made me feel like I didn't have the chops, the focus, to sit behind a desk in a damp office manipulating mathematical symbols for hours like the others. Fortunately, Chris had shown he believed in my abilities to make a contribution to cosmology by encouraging me to get out of the office and get more involved with my music. Working on physics ideas and calculations in between sets, at the jazz dives of Camden town, I found myself trying hard to believe that it would give me a creative edge in my research. That was a beginning. Ideas started flowing. But something more was about to change.

While Faye gave her living-room lecture, I honed in on someone else I had noticed throughout the evening. Dressed in black like Lee Smolin, he had a strong face and a gold tooth that shone every time someone engaged him in conversation. The way he listened to Faye, with such focus, I assumed he was a hardcore Russian theorist. It turned out he had come with Lee. When Lee noticed I was still hanging around after the talk, he invited me to join them as Lee walked his gold-toothed friend back to his studio in Notting Hill Gate. I was curious what research this friend was going to churn up and what school of quantum gravity he'd slot into. I had to work to keep pace with the animated duo as we walked along well-lit high streets, dipping in and out of dark London mews. This guy was no regular physicist, I soon realized. Their conversation was unprecedented. It started with the structure of spacetime and the relativity of time and space according to Einstein. That wasn't the strange part. Soon, they were throwing commentary around on the mathematics of waves and somehow kept coming back to music. This gold-toothed wonder was getting more intriguing by the minute.

That was my first encounter with Brian Eno. Once we reached his studio, we exchanged phone numbers, and he generously lent me one of his bikes—*indefinitely*. At the time, I didn't know who Brian was, but that changed a week later when I told a friend and band member about him. Tayeb, a gifted British-Algerian bassist and *ooud* player (an Arabic string instrument), was at first dumbfounded by my shameful ignorance. "Bloody hell, Stephon . . . you met the master."

Brian Eno, former member of the English rock band Roxy Music, established himself early on as a great innovator of music. He was part of the art rock and glam rock movement, when rock and roll took on a new sound by incorporating classical and avant-garde influences. The rocker look was dressed up with flamboyant clothes, funky hair, and bright makeup: think Lou Reed, Iggy Pop, and David Bowie. Brian was the band's synthesizer guru, with the power to program exquisite sounds. The beauty of synthesizers in those days lay in their complexity. In the early days, one had to program them—unlike synthesizers today, with preset sounds at the touch of a button. Popularity hit Roxy Music hard and fast, and Eno promptly had enough of it, so he left Roxy Music, and his career continued to flourish. He produced the Talking Heads and U2 and went on to collaborate with and produce greats such as Paul Simon, David Bowie, and Coldplay, to name a few. In addition, he continued with synthesizers and emerged as the world's leading programmer of the legendary Yamaha DX7 synthesizer.

I wondered why an artist like Brian would be interested in matters of space-time and relativity. The more I got to know Brian, I knew it wasn't a time filler, or for his health. What I was about to discover during my two years in London was that Brian was something I've come to call a "sound cosmologist." He was investigating the structure of the universe, not inspired by music, but with music. Often times he would make a comment in passing that would even impact my research in cosmology. We began meeting up regularly at Brian's studio in Notting Hill. It became a pit stop on my way to Imperial. We'd have a coffee and exchange ideas on cosmology and instrument design, or simply veg out and play some of Brian's favorite Marvin Gaye and Fela Kuti songs. His studio became the birthplace of my most creative ideas. Afterward, I'd head to Imperial, head buzzing, spirits high, motivated to continue my work on calculations or discussions on research and publications with fellow theorists.

One of the most memorable and influential moments in my physics research occurred one morning when I walked into Brian's studio. Normally, Brian was working on the details of a new tune—getting his bass sorted out just right for a track, getting a line just slightly behind the beat. He was a pioneer of ambient music and a prolific installation artist.

Eno described his work in the liner notes for his record, *Ambient 1: Music for Airports*: "Ambient music must be able to accommodate many levels of listening attention without enforcing one in particular; it must be as ignorable as it is interesting." What he sought was a music of tone and atmosphere, rather than music that demanded active listening. But creating an *easy* listening track is anything but easy, so he often had his head immersed in meticulous sound analysis.

That particular morning, Brian was manipulating waveforms on his computer with an intimacy that made it feel as if he were speaking Wavalian, some native tongue of sound waves. What struck me was that Brian was playing with, arguably, the most fundamental concept in the universe—the physics of vibration. To quantum physicists, particles are described by the physics of vibration. And to quantum cosmologists, vibrations of fundamental entities such as strings could possibly be the key to the physics of the entire universe. The quantum scales those strings play are, unfortunately, terribly intangible, both mentally and physically, but there it was in front of me—sound—a *tangible* manifestation of vibration. This was by no means a new link I was making, but it made me start to think about its effect on *my* research and the question Robert Brandenberger had put to me: How did structure in our universe form?

Sound is a vibration that pushes a medium, such as air or something solid, to create traveling waves of pressure. Different sounds create different vibrations, which in turn create different pressure waves. We can draw pictures of these waves, called waveforms. A key point in the physics of vibrations is that every wave has a measurable wavelength and height. With respect to sound, the wavelength dictates the pitch, high or low, and the height, or amplitude, describes the volume.

If something is measurable, such as the length and height of waves, then you can give it a number. If you can put a number to something, then you can add more than one of them together, just by adding numbers together. And that's what Brian was doing—adding up waveforms to get new ones. He was mixing simpler waveforms to make intricate sounds.

To physicists, this notion of adding up waves is known as the Fourier transform. It's an intuitive idea, clearly demonstrated by dropping stones in a pond. If you drop a stone in a pond, a circular wave of a definite frequency radiates from the point of contact. If you drop another stone nearby, a second circular wave radiates outward, and the waves from the two stones start to interfere with each other, creating a more complicated wave pattern. What is incredible about the Fourier idea is that *any* waveform can be constructed by adding waves of the simplest form together. These simple "pure waves" are ones that regularly repeat themselves, and we will discuss them in the next chapter.

Linked by the physics of vibration, Brian Eno and I bonded. I began to view Fourier transforms in physics from the perspective of a musician mixing sound, seeing them as an avenue for creativity. The bicycle Brian lent me became the wheels necessary to get my brain from one place to another faster. For months, the power of interdisciplinary thought was my adrenaline. Music was no longer just an inspiration, not just a way to flex my neural pathways, it was absolutely and profoundly complementary to my research. I was enthralled by the idea of decoding what I saw as the Rosetta stone of vibration—there was the known language of how waves create sound and music, which Eno was clearly skilled with, and then there was the unclear vibrational message of the quantum behavior in the early universe and how it has created large-scale structures. Waves and vibration make up the common thread, but the challenge was to link them in order to draw a clearer picture of how structure is formed and, ultimately, us.

Among the many projects Brian was working on at the time was one he called "generative music." In 1994 Brian launched Generative music to a studio full of baffled journalists and released the first Generative software piece at the same time. The Generative music idea that came to fruition about a decade later was an audible version of a moiré pattern. Recall our pond ripples interfering to create complex patterns. These are moiré patterns, created by overlapping identical repeating patterns, and there are an infinite variety of them. Instead of two pebbles creating waves, generative music rested on the idea of two beats, played back at different speeds. Allowed to play forward in time, simple beat inputs led to beautiful and impressive complexity—an unpredictable and endless landscape of audible patterns. It is "the idea that it's possible to think of a system or a set of rules which once set in motion will create music for you . . . music you've never heard before."[1] Brian's first experiment with moiré patterns was *Discreet Music*, which was released in 1975. It remains a big part of his longer ambient compositions such as *Lux*, a studio album released in 2012. Music becomes uncontrolled, unrepeatable, and unpredictable, very *unlike* classical music. The issue becomes which inputs you choose. What beats? What sounds?

What I began to see was a close link between the physics underlying the first moments of the cosmos—how an empty featureless universe matured to have the rich structures that we see today—and Brian's generative music. I began to wonder if structure could have originated from a single starting pattern of waves, like Brian's generative sound. I needed Fourier transforms and inspiration from Brian's musical brain. After all, he was playing with the Fourier idea with an intuition that transcended that of most physicists. I wanted to develop this intuition to be able to be creative with it. When I walked up to him as he was manipulating the waveforms that morning, he looked at me with a smile and said, "You see, Stephon, I'm trying to design a simple system that will generate an entire composition when activated." A lightbulb flickered in my brain. What if there were a vibrational pattern in the early universe capable of generating the current complex structure that we live in, the complex structures that we are? And what if these structures had an improvisational nature? There were some lessons in improvisation I first had to learn.

7

THRIVING ON A RIFF

During the summers, I would take a break from Imperial and visit Brian Greene's group at Columbia University's Institute for String Cosmology and Astrophysics (ISCAP) to work on a new project, striving to bring a key idea from string theory into cosmology. But at the time I had no idea that the connection would come from a chance meeting with a jazz legend. Although I eventually decided to take the job at Imperial College, Brian Greene was the first person to offer me a position, after a five-month stretch of postdoc rejection letters. Brian was known for his groundbreaking work on topology change in string theory, yet it was his passionate investigation into what string theory could say about the early universe that prompted my visit. The institute, formed in 2000 with Greene as codirector, was a natural evolution from his research program, applying superstring theory to cosmological questions. These programs have provided opportunities for many young cosmologists of my generation, for which they will be forever grateful. I remain indebted to Brian for making me that job offer. Fortunately, after I decided to go to Imperial, he extended to me a visiting postdoctoral position at ISCAP, so in the summer I'd travel from London to New York to visit ISCAP and do calculations and play at my favorite jazz spots.

But I wasn't the only expatriate New Yorker physicist who would return home for his homeboy fix. Lee Smolin was also in New York doing his own calculations on an exciting idea concerning dark energy and quantum gravity. We agreed to hook up to share our ideas and arranged to meet at the loft of one of his best friends, Jaron Lanier. Lee referred to Jaron as a genius. If Lee calls someone a genius, that is not to be taken lightly. I jumped on the 2 train from the Bronx to Tribeca and entered a huge loft, where, on one end of the loft, I saw literally hundreds of exotic instruments. On the other end were all sorts of electronic and computer devices. Lee greeted me and after a few minutes a tall, largish man decked in a pajama-like black T-shirt, black baggy pants, and sandals, with long, thick blond dreadlocks, emerged. He walked over and gave me a big teddy bear hug as if we had been friends for a while. This was the world-class computer scientist and composer Jaron Lanier, a pioneer of virtual reality. A quick glance across the room revealed the first *Wired* magazine cover featuring the dreadlocked Jaron, wearing goggles and gloves, as if he'd beamed down from some alien planet.

It didn't hurt that I also had long dreadlocks at the time. As the years rolled on, Jaron became one of my best friends. It's an understatement to say that Jaron is a polymath: artist, scientist, composer, multi-instrumentalist, and author. But what really amazed me about Jaron was his improvisational way of doing science and music. He'd take ideas from one branch of science and music and turn it into either a new piece of technology or scientific development. Like Mr. Kaplan, Jaron inspired me to take my hobby of connecting the worlds of physics and music seriously.

When we first met in 2000, Jaron mentioned that he was fascinated by a neural network of the visual system of the fruit fly. Because they reproduce rapidly, fruit flies are good test subjects for experimentation in genetics as well as neurobiology. As a result much biological information on the neural circuitry for fruit flies exists. Jaron and his colleagues were interested in making computer algorithms that simulated these neural codes. But at that time, I was not able to see the implications of the project and thought to myself, "So what?" Nine years later, Jaron bought a beautiful house on top of a mountain overlooking San Francisco Bay. I came out to visit, and then one day, on a walk through

Berkeley Hills, Jaron nonchalantly said, "Stephon, remember that fruit-fly thing? Well, some buddies of mine turned the technology into a start-up and wrote me a check. That's how I got this house." Not bad for someone who never went to high school. Well to be fair, Jaron did build and live in his own geodesic dome tent in New Mexico and attend college math classes as a teen.

Jaron also played the saxophone, so at some point during that first meeting in New York, sax talk came up. "You know, Stephon, my buddy Ornette Coleman lives uptown. How about we go visit him?" My jaw dropped. This is the same Ornette Coleman I had listened to as a kid in the Bronx, my first serendipitous exposure to jazz improvisation. Lee responded before I could stop daydreaming: "Oh, that would be fantastic." Jaron picked up the phone, and after a yellow cab ride, we found ourselves in Ornette's midtown palace.

Ornette Coleman was raised in the blues and folk tradition of Texas and is one of the major innovators of free jazz. At the time I was (and still am) studying what some musicians consider straight ahead or classic mainstream jazz. As with theoretical physics, it's necessary to first master a whole body of knowledge in order to play straight ahead. As an example, if you are in a jam session, and someone called a tune, say "Autumn Leaves," you would be expected to know the head (the beginning melody) and the remaining form (the harmonic and rhythmic structure). So an improvised solo on classic jazz is constrained by the structure or form of a song. But my discussions and lessons with Ornette would change the way I thought about improvisation and its connection to theoretical physics.

Ornette was a gentle, calm man who spoke often in parables. When I first met him, he took me to his studio and showed me his quintessential white alto saxophone. Then he gave it to me and said: "Give it a try." Imagine: one of the legends of jazz asking *you to play his horn*. On one level, it was beyond flattering; on another, very frightening. He gave me a clean mouthpiece and reed, and I started playing up and down a scale. Then, gently, he said, "It's amazing that there are only twelve notes, and you can have a conversation with those notes." It didn't take much for Ornette to inspire me. We started to talk more about his approach to improvisation—he's known for a new "attitude," or strategy, for improvisation that he calls harmolodics. In an interview he described the proccess:

> You take the basic notes and instead of approaching them in a restrictive manner and saying that you can't take that step if you play this note, start thinking about sound and what you can do with sound. That's really all I'm doing with harmolodic—thinking differently about melody, rhythm and harmony. It's more based on listening and responding, on sound and reaction, than any type of set pattern I could write out for you. Music has nothing to do with a lot of the things that people like to think it does.[1]

This was completely radical, given that I had always assumed that there was one right way to play jazz: Have all of your scales committed to memory and under your fingers. Practice hard and hone your technique so that you can play coherent and imaginative transitions through chord changes. Transcribe and analyze the solos of the greats for your particular instrument, which in my case meant John Coltrane, Sonny Rollins, Dexter Gordon, Charlie Parker, Wayne Shorter, and Miles Davis. But my first lesson with Ornette brought me back to some unexpected advice I got during my graduate school years. I was in the dark basement of a music store in Providence looking at old scores when I heard a raspy voice behind me. I turned to see a tall old man in a tweed jacket, who identified himself as a classical composer: "You're wasting your time. If you want to be a great musician, you need to know three things. First, you must master the rules before trying to break them. Second, music is about tension and resolve. Third, practice, practice, practice, but when you're out playing, forget it all." I never saw the man again, but his words stuck with me, and I often share this story with my physics students.

Like the fixed shape of a mountainous landscape, such as the one Jaron lives on, the structure of classic jazz songs provides a harmonic,

melodic, and rhythmic backbone for improvisation to unfold. For instance, many jazz tunes come from early Tin Pan Alley, Broadway, and
Hollywood songs, which jazz musicians use as base material to blow
and solo. The father of the tenor saxophone, Coleman Hawkins, was a
master of improvising on the *harmony* of a song. Lester "Pres" Young,
whose light, airy, yet intense style contrasted with Hawkins's rough-
edged approach, was a study in brilliant *melodic* improvisation. The
father of jazz, Louis Armstrong, and bebop genius Charlie Parker were
grand masters of improvisation on the melodic, harmonic, and *rhythmic* level.

But in harmolodics, Ornette would deliberately change the chords
in his improvisation. Unlike classic jazz where, generally speaking, harmonic movement throughout tonal centers guides the music, in harmolodics, melody, harmony, and *sound* all play an equal role in the
improvisation; like a symmetry principle, all of the elements of music are on the same footing. *Sound* is a difficult word to reduce to a
well-defined set of concepts; it is more of a metaphor, indicating how
each jazz musician has his or her own voice. Ornette Coleman and
Charlie Parker both play the alto sax, but each has his own sound or
signature—the distinct tonal quality and sonority, the way notes are
bent and placed rhythmically. A true jazz aficionado can hear a jazz solo
and identify who is playing.

When Ornette makes spontaneous changes, the band's reaction
would generate new structures in the music. Standard jazz guitarist
Marc Ribot observed how Ornette based these structures on motifs.

> Although they were freeing up certain structures of bebop, they were in
> fact each developing new structures of composition . . . The sets of rules
> for Ornette's harmolodic music . . . it's clear that it is based on taking
> motifs and freeing it up to become polytonal, melodically and rhyth-
> mically, it is tied very strongly to the motifs in the [main melody].[2]

A motif is a short melody that often repeats itself or recurs throughout a song. Perhaps the most famous motif is the beginning three notes

of Beethoven's fifth symphony. Throughout the symphony, the motif is recalled in different keys and played by various instruments. This approach was also reminiscent of Brian Eno's approach to generative music. They were both playing with the idea that structure and complexity can arise from simple rules or patterns. A serious listen to one of Ornette's songs will reveal that his solos are often created by modulating both his sound and the pitch.

After my short sax lesson, Ornette asked me what I was working on. I said vortices. Although vortices are commonplace in quantum field theory, I was thinking about them within the context of superstring theory (more on that later). Vortices are tube-like regions of trapped energy that are very common across nature. The swirly motion of water flowing down a sink is a vortex. Both a typhoon and the eye of a storm are vortices as well. Even in the quantum realm, magnetic fields can form a lattice of vortices in superconductors, which was a Nobel Prize–worthy idea. I got a piece of paper and drew the vortex for Ornette. He responded by saying that he plays vortex-like patterns in his solos. After this encounter, whenever I'd listened to Ornette's music it became clear to my ears that not only was he improvising notes but he was also forming geometric patterns, like vortices.

That encounter with Ornette influenced me years later when I coproduced my first album with electronic musician Rioux. The album, entitled *Here Comes Now,* was a tribute to Brian Eno and Ornette Coleman. It had elements of Brian's mastery of frequency modulation synthesis, and I played a lot of free jazz impressions over generative electronic rhythms. One of the best songs on the album, in my humble opinion, is Ornette's "Vortex."

Both Jaron and Ornette gave me a fresh way to view being a scientist and musician. Jaron was also a musician who effortlessly made useful analogies between music and science. I once saw him give a major computer science lecture in which he began his talk with an ancient Chinese instrument that he described as the first digital computer. Even though Ornette was not trained as a scientist, he would talk to me about my ideas in physics and how they could relate to music. One day

FIGURE 7.1. The cover of the 2014 critically acclaimed album the author coproduced: *Here Comes Now. Erin Rioux and Brandon Sanchez.*

he said to me, "I have a pattern to give you." He wrote down six notes on a piece of paper and said, "Get this under your fingers, and this will help you play through chord changes." Talk about relativity theory! Unfortunately I cannot reveal these secret six notes—yet.

As a young theoretical physicist, despite the encouragement of my mentors, I still felt the pressure to remain in the box and follow the herd. Making progress and moving up the ladder was largely based on whether you were respected by your peers. If you give any indication that you lack the chops of a well-trained theorist, you risk getting kicked out of the club. I was fully aware of my tendency to come up with ideas that some of my peers thought would get me rejected even from a crackpot asylum. And although I did everything in my power to master the traditional techniques expected of every theorist, I wanted to have in my head a picture of the physics. This was the same tendency in my music—getting a mental representation of my improvisations freed me from the formalism that I practiced and internalized. The initial meeting with Ornette ignited a breakthrough in my relationship with theoretical physics. I felt free and confident to stray from the herd.

Ornette took a big risk by breaking away from the bebop and classic jazz traditions. His love for new ideas gave him the courage to break the boundaries created by tradition. And this led to some really cool new music. In the same spirit, I realized that I could be a theorist who generates ideas for the sake of their beauty. I could manipulate these virtual theoretical worlds in the same way Ornette changed expected melodic paths constrained by Western harmony, for the sake of searching for a new idea and expressing what he was hearing. I realized many, if not most, of these speculative scientific propositions of mine would be wrong, but perhaps one or two might turn out to be an advance in my field.

Influenced by the years I'd spent listening to Ornette and his discussions with me, I refined my analogy between jazz music and cosmology. As Leon Cooper taught me, the best analogies can enable us to say something new about the physics that we otherwise would have not known.

8

THE UBIQUITY OF VIBRATION

Before the synthesizer and generative music, before the quantum cosmological debate of structure formation, there was Isaac Newton. With one intuitive and universal mechanism, Newton pulled together the works of three of his great mathematical predecessors—Pythagoras, Galileo, and Kepler. In 500 BCE, Pythagoras of Samos set out to reproduce the harmonics he heard in a blacksmith's shop. Attuning his ears to the music of hammers striking metal, he was able to reproduce the mathematical ratios of the hammer weights using taut strings of different lengths. But little did Pythagoras and thousands of years worth of investigators know that a universal secret underlying even the most complex vibrations would be encoded in a useful and elegant mathematical law—the Fourier transform.

Roughly two thousand years later, circa 1600, Galileo and Kepler maintained a passion for Pythagoras's discovery. Though they were unable to reach a physical understanding of how the strings produced harmonious tones, their life's work bore fruit that provided crucial stepping-stones toward this goal. Kepler's belief in divine geometry and cosmic harmony resulted in his laws governing planetary motion. What neither Galileo nor Kepler realized was that the movements they studied were direct manifestations of a single, yet unidentified force. In stepped Sir Isaac Newton.

Isaac Newton was born in England in 1642 and became one of the most influential physicists and mathematicians of all time. His talents were many, but his obsession was with motion. He helped invent calculus, contributed to optics, and, perhaps most importantly, laid the groundwork for classical mechanics. How objects on Earth moved, such as the movement of machinery, the motions of projectiles, and even Galileo's free-falling masses, was described by classical mechanics. Newton proved that objects move in response to forces, but his vision was further reaching still. He wanted to understand the motion of *all* objects, both on Earth *and* in space. His determination to do this resulted in the publication of his *Principia* in 1686, in which he identified an underlying force dictating both earthly and cosmic motions—gravity. Gravity caused objects to fall toward Earth and also held planets in orbit around the sun. His crowning success came when he used his new "universal laws" to *derive* Kepler's laws of planetary motion.

String harmonics remained elusive, but Newton, in discovering his laws of motion, unknowingly set the groundwork from which the physics of strings would finally be understood. Over time, his successors would fit together the rest of the puzzle pieces to complete the picture of string motion. From strings, a description of wave motion would follow, which, in turn, would ultimately prove to be the glue joining quantum physics, cosmology, and music.

The secret in Newton's discovery was in a phenomenon governing all objects in nature. It was the thread that weaved all motions together, and he called it the principle of inertia. Newton noticed that objects neither turn, speed up, nor slow down on their own. The *inertia* of an object, as he called it, is an inherent property that resists a change in motion.

Newton's first law of motion: Unless acted on by an external force, an object at rest stays at rest, and an object in motion stays in motion *with the same speed and direction.*

The first principle forces us to ask a new question: What is a force? To explain it, Newton formulated a second law. He defined "force" precisely by equating it with a change in velocity, where the velocity of an object specifies both its speed and direction. If an object changes its velocity—if it speeds up, slows down, or changes direction—it is said to accelerate. This is the essence of Newton's second law of motion. A force on an object causes that object to accelerate. Conversely, observing how an object accelerates helps determine the characteristics of the force. What Newton found was that the force acting on an object, and its acceleration, were directly related and generally depended on the mass of the object.

Newton's second law of motion: The force, F, acting on an object is equal to its mass, m, multiplied by its acceleration, a.

$$F = ma$$

Newton's insight into inertia, force, and changes in velocity may at first glance appear to be quite obvious, but it is very profound. This simple equation is able to dictate the position of a particle in the *future* when it is subjected to a force. Indeed, the magic of this equation lies in its power of prediction.

Newton's second law was the equation that my physics teacher, Mr. Kaplan, had written on the board the very first day of class. When he threw the tennis ball, his hand provided a force, and the ball accelerated upward. Gravity provided another force, and the ball accelerated downward. It would speed up, slow down, stop, and change direction—and $F = ma$ told the whole story.

To establish a visceral understanding of what acceleration is, let's put Newton in the driver's seat of a Formula One 1958 Maserati 250F. Imagine the delight that this man, obsessed with motion, would have had driving one of the nicest rides in history. Starting up the engine, Newton is still and the car's speed is zero, but as he steps on the accelerator, he can feel himself being pressed back into his seat. Newton nervously slams on the brake and jerks forward toward the steering wheel. After some practice, he gets the hang of it and cruises on a straightaway at a modest speed of 250 km/h. At a constant velocity, there is no acceleration, and our fascinated driver notices that there is no force pushing on him. Newton concludes that only a *change* in velocity causes him to feel a force, to be pressed back into his seat or forward toward the wheel. Similarly, changes in the car's direction, as the 250F hugs curves, cause a perceptible sideways force, which experienced drivers anticipate and lean away from.

Mathematically, "change" is normally represented by the symbol Δ. If we denote a position as X, then a change in position is ΔX. The fact that velocity is a change in position is thus written V = ΔX, and acceleration is A = ΔV = $\Delta\Delta$X = Δ^2X. Writing acceleration in terms of position means we can rewrite Newton's equation, F = ma, as F = m Δ^2X.

The predictive power of Newton's equation begins to emerge because it shows that a force on an object creates a definitive change in position of the object *over time*. Velocity is a change in position over time; acceleration is the change in velocity over time. So, to be accurate, position is written as a *function* of time, X(t). Similarly, velocity and acceleration are also functions of time, v(t) and a(t). To be precise, then, the equation should be written as a change with respect to time, and Newton's equation becomes

$$F = m\frac{\Delta V(t)}{\Delta t} = m\frac{\Delta^2 X(t)}{\Delta t^2}$$

Having Newton's equation in this form—with respect to acceleration, velocity, position, and time—a lot of information can be extracted.

Let's consider a series of four examples leading up to understanding the basic physics of strings.

Case 1: No external force. If there is no external force, F = 0. Assuming the mass of the object is nonzero, F = ma tells us that A = ΔV = 0. This is exactly Newton's law of inertia—if there is no force, an object's velocity remains constant. This is Newton cruising along the

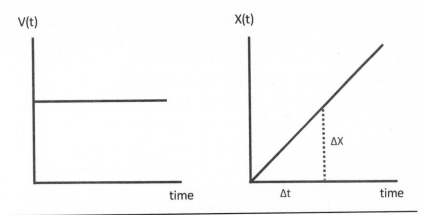

FIGURE 8.1. Graphs of the changes in the car's velocity and position.

straightaway. As the car moves, its position increases at a constant rate and can be described by a simple graph.

Graphs are a physicist's idea of fine art—an invaluable visual complement to an equation, revealing information about functions that may not be otherwise obvious. The graph itself can be read. The velocity graph in Figure 8.1, for instance, has zero slope and immediately tells us (with a bit of practice) that the function is unchanging in time—it is constant. The slope of the position graph, X(t), at any point is its rate of change, given by the value of v(t) at that point—a helpful visual tip-off! If the slope is steep, the rate of change is large, and vice versa.

Case 2: A constant force. Newton's equation for an object under the influence of a constant external force is.

$$F = constant = m\frac{\Delta^2 X(t)}{\Delta t^2}$$

If Newton, relying on his talent for precision, keeps his foot fixed on the gas pedal of the 250F without wavering, he would be applying a constant force. F = ma tells us that the car will accelerate at a constant rate. This phenomenon of constant acceleration due to a constant force is precisely what Galileo observed when dropping objects from the Tower of Pisa. Only there, the constant force was gravity!

Graphically, a(t) looks like v(t) before because it is constant, and v(t) looks like X(t) before because it is constantly increasing. The challenge is to figure out what X(t) will look like. This is a fascinating question because it will demonstrate the equation's predictive power—the position of the car at any time, t, under the influence of a constant force. We can use the graphs to get an idea using the fact that the slope of a function at a point is its rate of change. For instance, at t = 1, we can see that v = 1, and we know that the rate of change of x(t) is equal to v(t). So, at t = 1, the rate of change, or the slope of X is 1. Similarly, at t = 2, we have v = 2, and so the slope of x(t) is 2, and so on, ever increasing. Sketching out such a function gives us a shape like the one below.

The question remains of how to get a *precise* form for the function X(t). In working on this problem, Newton realized that he needed more mathematical umph. He needed something that would tell him how an object was moving at any given *moment* in time, not just over a

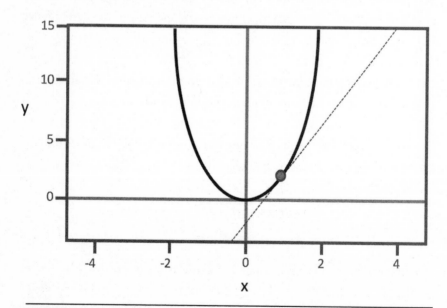

FIGURE 8.2. A parabola sketch of x(t) with slope lines at t = 1, 2, 3 would be better.

period of time. In Germany, Gottfried Leibniz, a man whose work has contributed to fields as varied as physics, linguistics, and politics, was thinking the same thing. Incredibly, in wanting to understand the motion of objects down to the smallest measurable changes, Newton and Leibniz both independently invented the branch of mathematics called calculus. It was the mathematics of change.

The old symbol was replaced with the *derivative* for an *instantaneous* rate of change, and amounted to sending the time interval, t, to the infinitesimal interval. With the derivative to hand, we have a new mathematical construct called a *differential equation*, and Newton's equation for an object under the influence of a constant external force becomes:

$$F = constant = m\frac{\partial^2 X(t)}{\partial t^2}.$$

Graphically, the first derivative of a function is its slope at a point, while the second derivative (as above) translates to the apparent curvature of the function at that point. The equation is asking us to find a function of position, $X(t)$, such that its second derivative, at all times, gives a constant. At this point, most mathematicians and physicists would venture to guess the form of $X(t)$ based on their previous knowledge of functions and their graphs. They would then "plug it in" to the equation to see if it works, tweak it accordingly, and voilà, a solution! In this case, it is a parabola, whose most basic form is given by $X(t) \sim t^2$. With practice, form and function become intuitively related. If we take the derivative, or slope, of $X(t)$ at any point, we will get the velocity function, with a constantly increasing slope. Taking the derivative of the velocity we get a constant acceleration, $a(t) =$ constant, consistent with a constant force.

The great significance of calculus is that one function can be *derived* from another via the mathematics of change, hence the name *derivative*. The derivative has become one of the most powerful tools in physics, engineering, and, as we will see, in acoustics as well. Making graphs essentially gives us the derivative without all the work because the form of the function—its shape, its apparent slope (rate of change), and its concavity, all related by derivatives—gives us information about the dynamics without even looking at the equations.

> Case 3: A nonconstant force. The classic example here is of a mass at-
> tached to a spring. Imagine pulling the mass slightly and releasing it.
> The mass will accelerate from zero. Now imagine stretching the spring
> much farther before releasing it. The mass will accelerate faster. As it
> turns out, it is a linear relationship, where the acceleration is propor-
> tional to the distance pulled, X. Newton's equation in its simplest form
> thus tells us that F α X. Newton's equation becomes:

$$\frac{\partial^2 X(t)}{\partial t^2} = \alpha X(t)$$

> where the constant of proportionality, incorporates mass, m, and the
> stiffness of the spring, k. The equation tells us that there is a function,
> X(t), whose second derivative gives back the original function.

The motion of a mass attached to a spring, whether hanging or laid flat on a frictionless surface, will be an oscillating motion, or a vibration, about a central equilibrium position. If you can imagine graphing this motion in time, you will find that it traces out a wave-like curve. The same occurs for any system whose force is proportional to the distance an object is displaced from its rest position, and all satisfy the above form of Newton's equation. The wave-like curve is a sinusoidal function, or sine for short, and is written x(t) = x(t) = sin (t). If we take two derivatives of a sinusoidal function, we get back the same sinusoidal function.[1]

The above equation describes, essentially, a single vibrating particle, which brings us one step closer to understanding the puzzle that stumped Pythagoras, Galileo, Kepler, and even Newton—the vibrations of strings. It also brings us one step closer to understanding sound waves and Brian Eno's synthesizer. Intuitively, we can see from the graph of the sinusoidal function that it could describe pure wave

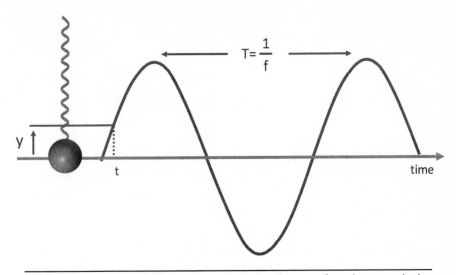

FIGURE 8.3. A sinusoidal function describing the vibration of an object attached to a spring about the equilibrium position.

motion, but let's try to understand exactly how by extending the example of a vibrating particle to the vibration of a continuous object.

Case 4: Another nonconstant force. Let's imagine plucking a guitar string and zooming in on a tiny piece of it. A simplified, yet accurate model of the string is to imagine the atoms that make up the string (microscopic individual masses) to be connected to each other by springs. In this uniform chain of atoms, each atom will independently oscillate up and down about an equilibrium position, just the way the mass attached to the spring did. Let's call this up and down distance "u."

Here, however, the individual masses tug on their neighbors, spatially distributed in the x-direction, via their attached springs, and something fantastic begins to happen.

As one mass tugs on the next and the next and the next, a wave travels down the string causing each mass, in turn, to oscillate up and down, the oscillatory motion passed from one particle to another—a traveling disturbance. Strings, of course, are not so fragmented as the

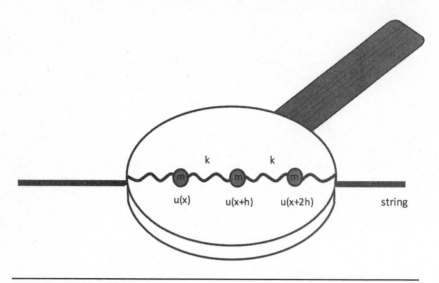

FIGURE 8.4. Magnification/string.

chain of masses we have described, so the derivative is used to shrink the distance between them so that we are describing, essentially, a continuous entity. Think of fans in a soccer stadium doing the wave. Viewed from afar, individual humans are barely discernible, and we see a human wave traverse the stadium. The distance away from the fans acts like a derivative, shrinking the distances between them down to nearly zero. So, in this case, Newton's equation describes the up and down motion of the entire string as a function of both time and position along the string, u(x,t), which takes the marvelous form:

$$v^2 \frac{\partial^2 u(x,t)}{\partial x^2} = \frac{\partial^2 u(x,t)}{\partial t^2}$$

Because there are two variables now, x and t, there are derivatives with respect to both. The equation tells its own story. It says the curvature of the string at a point (the second derivative with respect to x) will cause the string to accelerate at that point (the second derivative with respect to time). This is the equation describing the motion of a vibrating string. If only Pythagoras could enjoy this insight with us.

The solution for this equation is, again, a sinusoidal function, characterized by its height (amplitude) and wavelength (the distance from one crest to another). Sinusoidal functions come in two pure forms—a sine wave and its derivative the cosine wave, which is just a shifted sine wave. An amazing fact is that the sum of any number of sinusoidal functions gives back another sinusoidal function. The heights and wavelengths of the waves add up in such a way that they preserve the nature of the wave and thus remain solutions to the wave equation. This additive property of pure waves is the very Fourier idea that underlies Brian Eno's compositions and tells us that any shape that a vibrating string takes on can be gotten by adding up pure vibrations. This means that the equation for u(x,t) will not only describe a pure wave but any sum of waves.

Fourier idea: Any complex waveform that changes in time (like a complex sound wave) can be decomposed into pure sine waves of different frequencies and amplitudes.

In this sense pure harmonic sine waves can be used to custom-make any complex waveform—pure magic. As described in Chapter 6, when two stones are thrown into a pond, they create separate waveforms that eventually make contact with each other. The waves can interfere with each other either constructively or destructively. If the crests or high points of the waves coincide, they will add constructively, and the resulting wave will have a higher amplitude with the same frequency. But the waves can cancel each other out if the crest of one wave and the trough of the other coincide.

So, at the heart of the Fourier transform is the fact that waves interfere. The multitude of drops in a rainfall landing on a pond will create waves that interact with each other to create beautiful (or sometimes chaotic) patterns on the water's surface. We are now ready to mathematically express the Fourier idea. In words, the equation is:

A complex time evolving wave = A sum of sinusoidal waves

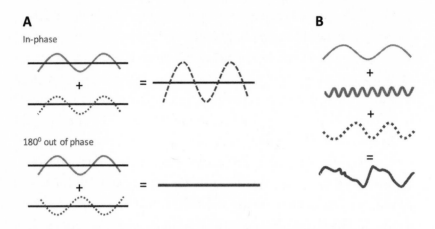

FIGURE 8.5. The diagram on the left (A) shows constructive and destructive wave interference. The diagram on the right (B) represents the Fourier idea by adding pure sine waves of differing frequencies, which add up to make a complex waveform.[2]

Because it can evolve in time, let's call the nontrivial (complicated) signal a function, F(t). The very powerful, beautiful, and ubiquitous Fourier *transform* is a mathematical equation that decomposes F(t) into its component waves, specified by their amplitudes, A, and frequencies. The equation then reads:[3]

$$F(t) = \sum_n A_n \sin(\omega_n t) = A_1 \sin(\omega_1 t) + A_2 \sin(\omega_2 t) + A_3 \sin(\omega_3 t) + \cdots.$$

We see in Figure 8.5B, the wave function, F(t), which is represented by the solid curve underneath the equal sign, arises from adding the above pure sine waves using the property that waves can constructively or destructively interfere; as seen in Figure 8.5A. We can see how electronic devices can exploit this idea to create electronic sounds out of oscillators, the heart behind the modern electronic synthesizer.

The Fourier transform is a mathematical operation that takes the complex function and decomposes it into its component pure waves by specifying the frequencies and amplitudes present. This is clear graphically. The *inverse Fourier transform* is the reverse mathematical

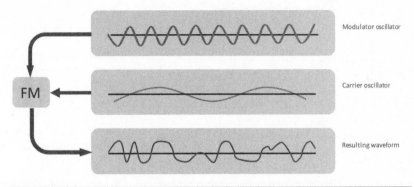

Modulator oscillator

Carrier oscillator

Resulting waveform

FIGURE 8.6. Frequency modulation synthesis that Brian Eno uses in his compositions.

operation, which takes the input amplitudes and frequencies and gives back the complex waveform as a function of time. The Fourier transform is one of the most used tools in all of physics, engineering, and computer science. It can be found in electronic circuits and lies at the foundation for sending and receiving signals to satellites and back to Earth through electromagnetic waves. And the Fourier transform is essential to understanding how the structure of the universe arises.

Now we have the tools necessary to understand the universal phenomenon called resonance. Sound, music, and many of the wonders of the quantum universe would not be possible without resonance. The physics of resonance governs why a sax plays a particular note and the condition for a particle to get created in a particle accelerator. In fact, it is one of the most widespread phenomena in all of physics. In short, resonance is the means through which vibrational energy can get transferred from one physical entity to another with great efficiency. Many objects, especially musical instruments (and, as we will see, quantum fields), are endowed with a natural frequency such that the disturbed object will oscillate at a unique frequency (or set of frequencies), depending on the properties of the material that the object is made of. The simplest example of natural frequency is a mass connected to a spring. The only two parameters are the mass and the stiffness of the spring.

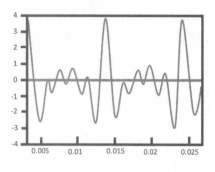 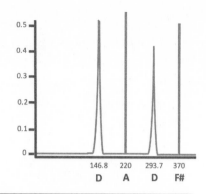

FIGURE 8.7. The signal to the left represents the amplitude versus time of the D-major chord. The curve to the right, the Fourier transform, shows the breakdown of the amplitudes and frequencies in a graph. Notice that in the figure on the right, only four frequencies are necessary to reconstruct the signal to the left. These are the four notes of the D-major chord.

Applying Newton's second law to a mass connected to a spring gives a mathematical relation for the system's "natural" frequency, $\omega = \sqrt{\frac{k}{m}}$. This immediately tells us that a stiffer spring will oscillate faster (because it has higher wave vector, k) and that a heavier mass (higher m) will make the system oscillate more slowly. If there is an external force that is pushing and pulling the mass at an arbitrary frequency unequal to the natural frequency, the spring will still oscillate, but its amplitude (the distance the mass traversed) will be less. But if the driving frequency matches the natural frequency, something remarkable happens. The amplitude of oscillation grows rapidly. This is the key to how musical instruments and even particle accelerators work.

Since strings can be thought of as a linear chain of many masses connected to springs, the string will have a larger set of resonant frequencies. We actually derived these sets of frequencies using the Fourier idea. In fact instruments are designed to resonate at a discrete set of frequencies corresponding to the notes of the musical scale. The key to musical instruments is to take a set of one or more driving frequencies (such as a vibrating reed, or the air flow from a flute) and control which frequency in the body of the instrument will resonate. In woodwinds, for example, this is done by closing a tone hole in the instrument.

Newton's law of motion unlocked the secrets of vibration and resonance from which, through the Fourier idea, we can both understand and construct complex waveforms from simple ones. As we will soon see, the Fourier idea applies to all the four forces and will serve as a key to understanding the structure of the universe. I leave you with a hint: if the structure of the universe is a result of a pattern of vibration, what causes the vibration? Is the universe behaving like an instrument?

9

THE DEFIANT PHYSICISTS

Theoretical physicist Jim Gates, one of my closest mentors and a pioneer in the theory of supergravity, which I was working on at Imperial, once told me that doing theoretical physics is like being a composer growing up on a planet with no sound and being challenged to compose a piece of music. This is how it can feel while trying to research the precise happenings of the universe fourteen billion years ago. Sometimes, the only thing that would help me gauge my progress and reground my footing in research was the healthy sense of rebellion and adventure I had learned from my PhD advisor Robert Brandenberger. Once in a while, turning around and embracing the new, I learned, was the secret to reviving the old.

Interestingly, quantum field theory, or QFT, caught my eye the day I "left" physics graduate school. It was one of those phases Gates had described: thinking about what caused the big bang was feeling a lot like trying to compose music on a planet with no sound, and I felt lost in it. I wanted something grounding, more connected to reality. In 1946, Erwin Schrödinger, the father of the Schrödinger wave equation central to quantum mechanics, wrote *What Is Life?*, a book that ignited the search for the role of quantum mechanics in living things. After reading the book, I became fascinated by the field of biophysics. After all, both nonliving and living things are based on molecules, and molecules are governed by quantum mechanics.

Nonliving matter can arise from atoms organizing themselves in a periodic repetition, like the periodic electron spins in the Ising model that give rise to different forms of magnetism. Schrödinger ingeniously concluded that the genetic code underlying life should be a quasi-periodic structure, like a helix. Looking down the vertical axis of a helix, you would see a circle, which is periodic, like a wave. But looking from the side, the periodicity is lost. Watson, Crick, and Wilkins were inspired to look for the double helix structure in DNA, found it, and shared the Nobel prize for the discovery. There is a fascinating connection between structure and formation here—the helical structure enforces the mechanical stability necessary to store the genetic material during the lifetime of the organism.

With that, my mind was made up to quit physics. I had to tell someone. Gerry Guralnik is best known for being one of the discoverers of the Higgs boson. With a teddy bear demeanor and a midwestern accent, Gerry was unusually approachable for a physicist of his stature and worried about our success. Like one of his mentors, Julian Schwinger,

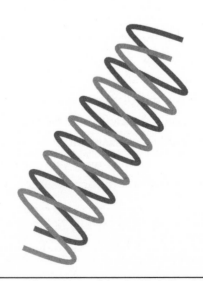

FIGURE 9.1. Quasi-periodic structure of the double helix of DNA.

he loved driving fast cars and making small talk over a pint of beer. His lectures had almost perfect attendance because he told great stories.

"Surely, Stephon, it's fine to quit physics . . . but don't quit science," Gerry said, with a look of genuine concern. "What else are you passionate about?"

I told him I wanted to understand the quantum origin of life. After a moment's contemplation, he said, "I got an idea. Did I ever tell you that I missed out on the Nobel Prize in chemistry?" He's a theoretical particle physicist, so I was sure he was kidding. As it turns out, his old PhD advisor, Wally Gilbert, had gotten involved in a molecular biology question with Watson and wanted him to join in. Wally left particle physics, a newly devoted biologist, and wanted Gerry to join him. "No way I was going to do that!" Gerry said. "But who would have thought he was going to win the Nobel Prize for his work in the genetic sequencing of DNA?! I'll tell you what, lemme call Wally now."

He picked up the phone, dialed; it rang. "Hi, Wally, it's Gerry. I got a student that you should meet. He's considering biophysics. When I last checked, you were the chair of Harvard's program." I hear the murmur of a voice on the other end of the line. "OK. I'll have him come see you next week."

The following week, Wally spent three generous hours telling me about his journey as a scientist. He clearly empathized with my situation and found me a job at the Harvard Medical School to work in X-ray crystallography, which is used to determine the three-dimensional atomic structure in viruses. It was time to pack up my books and head out of the Brown graduate physics department.

I was ready to move to the "greener pastures" of the marbled Harvard medical school halls. The physics grad students at Brown were used to small spaces filled with the funky smell of unshowered students and strong coffee. Small windows, late nights, and lots of problem sets were the norm. Years later, we weren't exactly surprised when we caught a rumor that a prison architect had designed the physics building.

As I walked out of room 122 of the Barus and Holley Building with a box full of books, I noticed a textbook, which I never found again, called, in part, *Quantum Field Theory* on the desk of a classmate. He wasn't around, so I snuck a peek. I recall the preface began: "Quantum field theory, the unification between special relativity and quantum mechanics, states that all matter and its interactions are composed of harmonious vibrations of fields. One is left with the vision that the entire universe is a symphonic orchestra of these fields."

How could I discover this gem the day I quit physics?! I could almost *feel* new neural connections forming as my brain ticked with excitement. The box was heavy and I left, but the image remained with me. I'd been baited.

I spent a challenging summer in the Hogle lab avoiding fatal chemicals. Jim Hogle, my supervisor, had been the one to figure out the three-dimensional atomic structure of the first animal virus, poliovirus, using the symmetry principles he had learned in group-theory classes at the University of Wisconsin. He was ingenious, with a history of success, but nonetheless would always welcome my speculative ideas. The man made an impact on me. One day Hogle casually, but seriously said, "You know, Stephon, I am keen on physicists trying to make a contribution to biology, but physicists have to respect the complexity in biological systems. The world is not made of spherical cows." He was gently making fun of all the simplifications physicists make in trying to understand the complex. Finding symmetries in equations is a grand method, and indeed a spherical cow is way easier to put on paper than the irregular real thing. Imagine living on the surface of a perfect sphere. No matter where you are on the surface, it will look the same.

Before my rendezvous with biology, I thought that the power of symmetry was exclusive to physics, but I was wrong. I learned that viruses were endowed with varying amounts of symmetry. Like Lego building blocks, the individual proteins self-assemble into an icosahedral structure characteristic of a virus. According to my friend Brandon Ogbunu, an MIT biophysicist, in biological systems symmetry can persist from the molecular to the organism itself to maximize its evolutionary fitness. A straightforward example is the bilateral symmetry of our legs—they have to be the same length if we want to run and hunt successfully

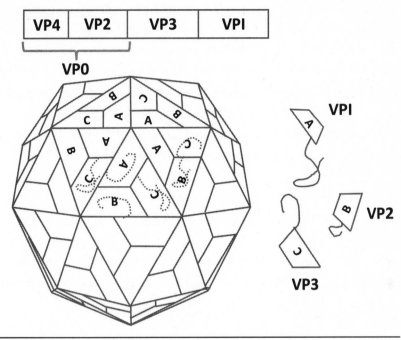

FIGURE 9.2. The icosahedral symmetry of the poliovirus. The subunits labeled by the letters A, B, and C are proteins that organize into triangular capsids. Twenty triangular capsids meet at twelve vertices, giving the three-dimensional icosahedral symmetry.[1]

in the jungle. The various symmetries of viruses function, for instance, to provide mechanical stability and the efficiency to bind to a host cell. What's more, viruses are made of hundreds of thousands of atoms, and knowing their precise positions in three-dimensional space makes the structure determination more difficult. Their structure reminds us that viruses and their constituent molecules are a quantum phenomenon.

Symmetry seemed like a key link between particle physics and the functions of life, yet lurking in the background of my mind was the whole reductionist debate in the sciences. The debate was succinctly and vigorously brought to light in a short article "More Is Different" by Nobel laureate Phil Anderson, which I read while working in the Hogle lab. It was about symmetry and fundamental physics. When particle physicists probed shorter distances and higher energies, they

discovered new symmetries, which simplified fundamental particle interactions. The underlying framework is described by quantum field
theory. In contrast, Anderson argued that "[t]he more the elementary
particle physicists tell us about the nature of the fundamental laws, the
less relevant they seem to the very real problems of the rest of science,
much less to those of society." He was making a key point—the high
degree of symmetry found in elementary particle physics is no longer
functional on the scales where complex phenomena occur:

> The behavior of large and complex aggregates of elementary particles,
> it turns out, is not to be understood in terms of a simple extrapolation
> of the properties of a few particles. Instead, at each level of complexity
> entirely new properties appear, and the understanding of the new be-
> haviors requires research which I think is as fundamental in its nature
> as any other.[2]

In other words, biological systems, such as Jim's viruses, are made up
of the same fundamental particles as those described by quantum field
theory, but the whole is not the sum of the parts, as the reduction of
symmetry is inherent with large numbers of atoms and molecules, with
increasing complexity. One question remained. Even though complex
viruses possess less symmetry, they still have symmetry. It was the interplay between symmetry and symmetry breaking that seemed to persist
across seemingly unrelated phenomena, from elementary particles, to
life, to the cosmos itself—as well as music.

Hogle made it clear to me that biology needs the foundations of
physics to be more developed because, one day, biology will evolve
enough to tackle questions like the role of quantum mechanics in the
functioning of viruses. "We're not there yet," he admitted. It was his
gracious way of saying that I wasn't cut out to be a lab biochemist, and
in hindsight and with gratitude, he was right.

Happily, Gerry was glad to see me return. I returned back to grad
school invigorated to understand why quantum field theory was the
fundamental language of physics and what symmetry and symmetry

breaking had to do with this. Maybe one day I'd learn how life emerges. Robert Brandenberger was already on the case. With his segue into Alan Guth's theory of cosmic inflation, he had clued in to how the quantum realm could provide the seeds for current large-scale structures in the universe. It wasn't quite life, but life needs planets and stars to exist. Robert had found renewed interest in QFT, and now so had I, but through biophysics. As a well-trained physicist, I had always been convinced that beauty and elegance rested in symmetry, but biology had taught me that there is something deeply profound and beautiful in *broken* symmetries. Robert and I were on a quest to understand the subtle asymmetries in the early universe that led, ultimately, to us.

I had taken a risk. It had been a bit like realizing, in the middle of a jazz solo, that the moment called for a deliberate wrong note. As one of my jazz teachers once said, "You practice all these scales, exercises, and long tones so that in the middle of a solo, when you play that wrong note, like an acrobat, you'll know how to fall." I had decided to play some wrong notes in my research, fell, and learned a hell of a lot in the process.

10

THE SPACE WE LIVE IN

At the same time that I was beginning to think about what Brian Eno, the sound cosmologist, was showing me about the structure of the universe, I was also still working on the problem the old-fashioned way: with Einstein's general theory of relativity. Ever since that day in Mr. Kaplan's office when he told me about Einstein's groundbreaking theory of space-time, I had been on a journey to master it. I finally did master it some two decades later as a graduate student when I was finally able to manipulate Einstein's field equations to understand the space-time structure of the universe. This was very different from the rote problem-set solving and examinations of graduate school course-work that I had endured. Now I could riff off the Einstein field equations, just like I did with my sax with every newly mastered scale. I was a cosmologist, playing with the universe, playing with space and time, and the matter within it.

The space out there is the same space that separates you from your book. For centuries philosophers and astronomers have assumed that space is empty, an inert medium in which the real stuff, like matter, moves about. But after thousands of years of brilliant philosophers getting it wrong, Einstein—whose genius partly lay in his courage to question the basic assumptions in accepted physical ideas—showed us that space is far more interesting than even the matter that moves within it.

The first idea Einstein questioned was gravity. Galileo's experiment at Pisa had proved that two falling balls of differing masses accelerate at the same rate. Einstein took this experiment beyond Earth and the solar system and, in so doing, changed the Newtonian descriptions of gravity and motion forever. He began with a thought experiment. Here is a modern version of it.

Consider one person inside a rocket ship at rest on Earth, and another person in a rocket ship in outer space. The person on Earth experiences the planet's gravitational pull and has the perception of not moving.[1] For the person in outer space, as long as the rocket ship isn't moving, that person will float because there's no gravity. However, if the rocket accelerates, then the person will have the impression that he has weight, because he will be pushed against the floor of the rocket. Einstein concluded that there is no way for either person to discern whether he or she is at rest in the presence of a constant gravitational field or accelerating in empty space. Both situations, he decided, are physically equivalent, depending only on relative states of motion. This "equivalence principle" is the crux of Einstein's general theory of relativity.

The consequences of this simple, almost childlike thought brought one of the most beautiful branches of mathematics, differential geometry, to the forefront of gravitational physics. Differential geometry describes a coordinate system. Einstein generalized his theory—thus creating the general theory of relativity—by animating the coordinate system itself. The structure of space and time itself, he argued, is dictated by the structure of matter. He unified space and time into a single coordinate entity, space-time, and described how it bends in the presence of matter and energy and how, in turn, matter moves according to where space-time is warped. As physics great John Archibald Wheeler succinctly said, "Matter tells space-time how to curve, and curved space-time tells matter how to move."

So in the case of Einstein's rocket ships, one passenger experiences acceleration because the Earth bends space, causing a gravitational force. For the other, the applied energy from the rocket boosters bends space, which in turn accelerates the rocket.

At the time Einstein was working on his thought experiments, one of the great puzzles in physics was the orbit of Mercury around the sun, which didn't follow the path predicted by Newton's theory of gravity. In 1915 Einstein worked out that his theory of curved space-time accounted for Mercury's abnormal revolution around the sun. Mercury is so close to the sun that its Keplerian orbit is altered by the sun's massive gravitational effect. Einstein was convinced that its orbit reveals the way the sun's gravity warps space and time around it. But in 1919 the life of *all* cosmologists changed when a second of Einstein's predictions was vindicated. Einstein had predicted that during a solar eclipse a star situated behind the sun would be visible because its light would follow a curved path *around* the sun. His theory was correct. But this was just the tip of the iceberg. Beyond our solar system, Einstein's equations could be used, amazingly, to describe the space-time of our entire universe.

As beautiful and awe inspiring as general relativity is, it is an extremely difficult theory to work with in practice. General relativity provides equations for both the motion of objects as well as the curvature of the space-time field of gravity, which makes it difficult to find exact solutions. Unlike Newton's theory of gravity, which is defined by one equation, the general theory of relativity has *ten* interdependent differential equations that need to be concurrently solved. But this did not stop Einstein and his contemporaries from searching for solutions.

His theory worked well for the solar system and resolved the anomalous motion of Mercury, but when it came to the universe, Einstein was perplexed. His theory predicted that the universe had to expand. But observations up to that point in time had shown that the universe was static. Einstein, clever as ever, "fixed" the expansion problem by introducing a term into his field equations, called the cosmological constant, that could be tuned to halt expansion.

Then in 1927, astronomer Edwin Hubble showed Einstein his data. Einstein realized that in introducing the cosmological constant he had made his biggest mistake. For the first time in history, Hubble, by taking pictures of galaxies, was able to calculate their velocities and distances. If the universe was static, all galaxies would have the same velocity regardless of their location. To everyone's surprise, especially Einstein's, the results showed that all galaxies move faster the farther apart they are. Einstein immediately knew that this meant the universe was expanding.

As it turns out, expansion is favorable to finding solutions to Einstein's equations because it allowed the same principle that Copernicus applied to our solar system circa 1500—that we are *not* the center of the universe—to be applied to the universe. In so doing, four physicists independently found exact solutions using Einstein's theory that described a perfectly symmetric expanding space.

To see the Copernican principle at work, we need to go back in time. Since the universe is expanding, we can, theoretically, run the universe's clock backward and shrink it. As the universe contracts, the matter in stars, planets, and galaxies gets condensed into a smaller and smaller space. If we run the clock back far enough, the atoms in all this matter begin to change. On our usual scale, at low energies, electrons bind to the nucleus of an atom. But at high densities, thermal energies surge and knock electrons out of their orbits. This means that right after the big bang, the early universe was filled with hot, dense free electrons, nuclei, and photons. In its infancy, our universe was a featureless distribution of excited matter and radiation. It was a seething plasma with no structure, a "primeval fireball"—a Copernican cosmos. This may not sound like a terribly interesting universe, but it is at least one for which Einstein's equations can be used to solved it. This vision of the early universe raised the question that ultimately I would work on: What was responsible for converting that incipient plasma into the stars, galaxies, and planets we see when we look up into the night sky?

Some people might think that a good physical theory should be flawless, especially some of my colleagues in search of a theory of everything. I don't believe we will ever have that perfect theory. Nature, like a great improviser, will always come up with new surprises that our theories will fall short of explaining or predicting. Plus, a good physical theory usually points to the roots of its own destruction. This is, actually, the case in Einstein's expanding universe hypothesis. The hypothesis makes beautiful predictions of the observed ratios of light elements in galaxies and Hubble's law of receding galaxies, but it alone can't tell us what drove structure formation. The trick is to preserve the accurate predictions of the theory, yet liberate the theory from its failings. Let's find the cosmic perpetrator.

Perhaps the most important prediction of the expanding universe comes from the period when the first elements were formed. As the hot electrons, which were zipping around, became more dispersed, they cooled, and their motion became less frenetic. Protons were lying in wait to capture them, and the first light element, hydrogen, was about to be born. The conditions for hydrogen to form occurred about 380,000 years after the big bang, when the universe had cooled to 3,000 degrees Kelvin, a balmy 5,000 degrees Fahrenheit. At this temperature, the energy of the particles was low enough for the Coulomb force, the attraction between oppositely charged protons and electrons, to take over and link them up to form hydrogen. But most of these electrons were still highly energetic, making the hydrogen unstable. To form *stable* hydrogen, the electrons had to fall to the lowest possible energy level, which meant that the excess energy was released in the form of photons with a characteristic temperature of 3,000 degrees Kelvin. The universe literally glowed.

These predictions were first made in 1948 by George Gamow, Ralph Alpher, and Robert Herman. Einstein's equations were saying something precise about the universe billions of years in the past. It was truly awe inspiring. But even more inspiring would be to find proof of this epoch, by finding whatever remained of the glow of the early universe. The hunt would be a major test for the big bang paradigm as well as the prediction that the early, plasma-filled universe was as uniform—and as Copernican—as everyone thought.

For years, cosmologists looked for the relic radiation that was expected to pervade all of space. As the universe expanded, the light waves from that time would have stretched to one thousand times longer than their original wavelength. That would result in a set of waves permeating the universe with wavelengths corresponding to microwave photons, like a microwave oven. In 1967, the cosmic microwave background, or CMB, was discovered at Bell Laboratory in New Jersey "by mistake" by engineers Arno Penzias and Robert Woodrow Wilson and won them the Nobel Prize in physics. Working on a telescope built to detect discrete radio wave signals, Penzias and Wilson were hampered by a persistent source of interference. Working to eliminate it, they isolated their telescope from a range confounding inputs, including other radio waves, heat from the mechanism itself, and even, in desperation, pigeon droppings. Even so, they could not eliminate one particular background hum, which came uniformly from everywhere. This, they ultimately reasoned, was beyond their mechanism, beyond Earth itself. It happened to be the very thing that cosmologists Robert Dickie, Jim Peebles, and David Wilkinson were gearing up to look for at Princeton University, just down the road. The Princeton group had built an instrument, the Dickie radiometer, to look for such background radiation. Importantly, they also had the knowledge to evaluate it.

There it was, the CMB, in all its subtle glory, surrounding us all the time, an imprint of the formation of the first stable atoms. The featureless CMB photons surely confirmed the cosmological principle and the expanding universe paradigm, but there was a serious problem lurking in this fossil imprint of the early universe. The discovery of that problem would open a Pandora's box, revealing the flaw in the Einsteinian vision of the expanding universe.

Standard big bang cosmology predicts that the particles in CMB light radiation will all have the same temperature. However, if every particle in a gas has nearly the same temperature, it means that they had to have interacted with each other at some time in order to attain such thermal equilibrium, such uniformity. They needed to have been in causal contact in the past. Consider CMB radiation coming at us from two opposing directions. Electromagnetic radiation—whether visible light, radio waves, or microwaves—travels at the speed of light, the maximum speed that physics allows. We can rewind the universe's expansion and watch as the radiation travels at the speed of light backward in time to 380,000 years after the big bang. The radiation from one direction is traced back to one particular region of the universe, while that from the opposite direction is traced back to a different region. This covers the time it took for both regions to reach *us*, but for *those* two regions to have been in contact with one another, it would have taken longer since they are in opposite directions from us. Given what we know about the expansion rate of the universe, measured from the recession velocities of distant galaxies, a disturbing conclusion emerges. It would take longer than the lifetime of the universe for these two regions of the CMB radiation to have been in causal contact. This is known as the "horizon problem." The very success of the big bang paradigm, the prediction of the thermal equilibrium observed in the CMB plasma, points to its own destruction.

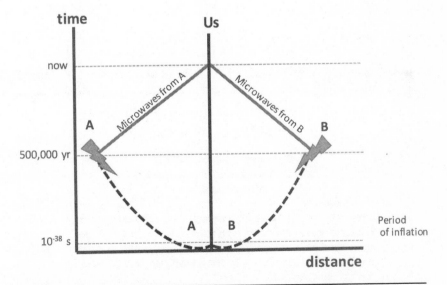

FIGURE 10.1. The regions A and B represent microwave radiation with the same temperature from regions that were not able to interact with each other. The big bang model lacks a causal means for these regions to reach thermal equilibrium, unless they had longer than the lifetime of the universe to do so.

Soon after the discovery of the CMB, a young graduate student, Bruce Partridge, and his professor, David Wilkinson, built a detector to see whether the radiation from 380,000 years after the big bang was as featureless as Copernican uniformity would imply. They were hoping to find irregularities, called the CMB anisotropy, to understand where the irregular structures, such as star clusters and galaxies, come from.

The idea was that if there existed tiny undulations in the primeval fireball, they would have grown with the expansion of the universe into large density fluctuations—variations that would cause gravitational instability and hence the beginning of the gravitational collapse of matter into structures. It was a beautiful theory: anisotropies as the seeds that gave birth to large-scale structure. Finding them would shed light on how the universe evolved from a Copernican beginning to our current, and rather different, cosmos. Unfortunately for Partridge and Wilkinson, their detector wasn't capable of measuring any anisotropy. The hunt, however, went on.

VOLUME 18, NUMBER 14   PHYSICAL REVIEW LETTERS   3 APRIL 1967

ISOTROPY AND HOMOGENEITY OF THE UNIVERSE FROM MEASUREMENTS
OF THE COSMIC MICROWAVE BACKGROUND*

R. B. Partridge and David T. Wilkinson†
Palmer Physical Laboratory, Princeton, New Jersey
(Received 2 March 1967)

A Dicke radiometer (3.2-cm wavelength) was used to make daily scans near the celestial equator to look for possible anisotropy in the cosmic blackbody radiation. After about one year of intermittent operation we find no 24-h asymmetry with an amplitude greater than ±0.1% (of 3°K). There is, however, a possibly significant 12-h anisotropy with an amplitude of about 0.2%.

FIGURE 10.2. The abstract of Partridge's and Wilkinson's article published in *Physical Review Letters* (April 3, 1967) of the first attempt to detect the intrinsic anisotropy of the CMB.

I learned about these problems many years later. By that point, Partridge—or Bruce, as his students used to call him—was a professor at Haverford College in Pennsylvania. He was a gentleman in every respect, famous among students for his crystal-clear, formal, and highly organized lectures. With his generous warmth, he was accessible to even the most timid and intimidated students. I was one of them. Like me, Wilkinson had been a saxophone player before becoming a cosmologist, but it was Bruce, in the end, who played a major role in my undergraduate years. During my sophomore year, Bruce had his colleague from MIT, Alan Guth, come to Haverford to give a talk to our class. In that season of my life, physics was an ongoing interest of mine, but I was a little defiant. Wearing African medallions and Malcolm X T-shirts, my hair in dreadlocks, I'd sit at the back of the classroom and rarely participate, my headphones tuned to the pro-black raps of Public Enemy. It was 1990, and a NASA satellite had an experiment on board to look for the anisotropies in the CMB radiation that Bruce and Wilkinson first attempted to find in 1967.

Alan Guth was the inventor of cosmic inflation, which both provided a solution to the horizon problem and insights into the anisotropy question. If the universe went through an exponential phase of growth in its infancy, Alan proposed that the distance radiation traveled would be boosted by the rapid expansion. This could explain the causal connection between regions that had previously seemed unable to communicate because the age of the universe did not leave them enough time to do it. This was a megatheory, in accord with CMB observations. Furthermore, because Alan's theory of inflation had a quantum nature, it allowed him to make a prediction about the nature and origin of the postulated CMB anisotropies. This quantum aspect of the early universe was what Robert Brandenberger was so keen to explore, and little did I know that inflation was going to be a major part of my future research.

Maybe my headphones had been on too tight, but at that time, I was unaware of the significance of Alan's visit to Haverford. I knew he was important, though, and was ready to give my rebellion a little rest.

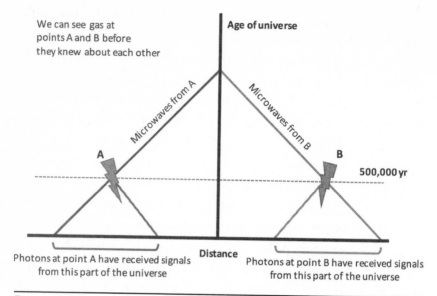

We can see gas at points A and B before they knew about each other

Age of universe

Microwaves from A

Microwaves from B

A

B

500,000 yr

Distance

Photons at point A have received signals from this part of the universe

Photons at point B have received signals from this part of the universe

Figure 10.3. Regions A and B represent microwave radiation with the same temperature from regions that, in the standard big bang model, were not able to interact with each other. A period of exponential expansion, or inflation, in the early universe provides the causal connection needed to explain the uniformity of the CMB.

After all, this was the first time I had heard about Einstein's theory at work since I'd spent time with Kaplan. This was beyond the mundane physics and mathematics I was learning as a sophomore. This was cosmology. This was the evolution of the universe as a whole and how it unfurled, reflecting Einstein's mathematical brilliance. The CMB was grand, the theory of structure formation equally as grand, and now this idea of inflation. It was hard to swallow.

Famous cosmologists from Penn and Princeton came to our classroom to hear Alan speak about inflation, which left the students too intimidated to ask questions. Nevertheless, after the talk, Bruce said, "First we'll get some questions from the students." My hand made a tremble to rise before being instinctively restrained. "Stephon, I see that you have a question," Bruce quickly said. He knew me too well. I had that sinking feeling that comes with a mixture of stupidity and naïveté.

The words spilled out before my conscious mind had a chance to inter-fere. "Does inflation do work?" In Bruce's introductory class we learned that work is done whenever force is applied to move an object over a distance. Since inflation is making the universe expand, I was confused as to what force was causing the universe to inflate. Is it possible for the universe to expand without any work being done? The wonderful thing about physics are these moments when the "rules" that we think as unchangeable are broken. I wanted to know. Alan responded, "That's a great question . . . inflation works to ignite the expansion of the uni-verse. We call the agent that does this work the *inflaton field*." Bruce had no idea the positive impact that his prodding and Alan's seriousness in his response to my question had on me. Even today, I beg my stu-dents to ask me "dumb questions," as they are usually the difficult ones.

Finally, three decades after Bruce and Wilkinson first looked for CMB anisotropies, the Cosmic Background Explorer (COBE) satellite directly measured it. After four years of space-borne probing of the CMB, an instrument on board COBE, the differential microwave radiometer, detected the feeble variation. At last, after all those years searching for clues about our origins, cosmologists finally entered the golden era of precision science. The discovery made inflation all the more important: without it, those perfect imperfections compounded the horizon men-ace because it meant that an explanation was needed for both the nearly perfect thermal radiation, plus the tiny undulations in the radiation's pattern. Inflation could provide a solution here, too. But there was more.

Every discovery seemed to unearth deep questions. The CMB sea was confirmed. The anisotropies within it as well. And all the while, bit by bit, astronomers were meticulously mapping the largest struc-tures in our universe. The Hubble space telescope became Earth's eye to the universe. It photographed glorious and variable objects in our vicinity, like nebulae and colliding galaxies. But technological advances allowed for different perspectives. Telescopes sensitive to radio signals, microwave signals, and infrared and gamma rays could all produce im-ages alone or in collaboration. Zooming out to the largest observable distances, these telescopes provided a map of the observable universe and revealed a surprise. Like Geller and Huchra showed with their map of large-scale structure, clusters of galaxies align and clump into walls and filaments. The largest formations in our universe, it turns out, have a homogeneous and isotropic distribution. This is formally known as the cosmological principle. So many years of scientific inquiry, technological advancement, and human ingenuity have unearthed that there are many levels of structure permeating our universe.

The origin of the hierarchy of structure in our universe remains elusive, but inflation provided a huge step in understanding how quantum fluctuations in the early universe could create asymmetries in the distribution of the otherwise uniform primeval fireball and be magnified by the inflationary expansion of space-time. Subsequent observations were to unearth harmonies in the early universe anisotropies, which couldn't have been created without the existence of a cosmic horizon. To better understand horizons, let's turn to sound.

11

SONIC BLACK HOLE

Lurking in every active galaxy in our cosmic web is the densest and most elusive object known to physics, the black hole. It was one of the very first exactly solvable systems in general relativity and at first was thought to be a purely theoretical construct. But black holes are cloaked by a horizon, similar to the cosmic horizon in cosmology. By exploring the role of horizons surprisingly to sound, in the case of black holes, we will gain a deeper intuition in our search to link music with cosmic structure.

The difficulties in dealing with Einstein's ten coupled equations, versus Newton's one equation, are enormous. Imagine a series of masses on springs linked together and in motion. Newton's differential equation can be applied to the movement of one independent mass. But because the masses are linked, the motion of one will influence the motion of those it is attached to and hence affect *their* equations of motion. A set of coupled differential equations is needed to determine their overall motion. To solve one equation, you are required to solve them all. A similar situation arose in the Ising model of magnetism discussed in Chapter 2, where the spins of neighboring atoms influenced each other and hence the overall interaction energy of the system. The difficulties are compounded when we recall that Einstein's equations couple not just mass with mass but mass with space.

To really understand the magic behind Einstein's ten coupled differential equations, it is useful to begin by considering a solution to them. But given their complexity, it is no easy task to dream up a physical space-time configuration that satisfies them. It's no longer a case of studying a graph and guessing the form of the function, as we did with Newton's equations. Even today, with the help of powerful computers, we still cannot find exact solutions of the gravitational field for interesting astrophysical systems. Nevertheless, just after Einstein developed his theory, physicists were buzzing with curiosity about his new space-time concept and eager to find solutions. For starters, they armed themselves with Dirac's trusty method: using the power of symmetry.

The great power of mathematical symmetry is that it can reduce the complexity of the equations. Imagine there are two separate equations that describe the oscillation of two particles, particle X and particle Y. One example of a "symmetric" situation would be if the behavior of X was exactly the same as Y. The two differential equations could thus be reduced to one, and once a solution for either X or Y was found, the solution of the other would follow.

Sometimes, nature actually provides these serendipitous situations of high symmetry, and physicists can delight in discovering the solutions. In the case of Einstein's equations, spherical symmetry was a good place to start. Spheres could model the structure of stars, like our sun. The geometry of spheres allowed gravity to be reduced to a radially uniform field around a compact central source. It was such a natural and simple idea that, within a few months of Einstein developing his theory, Karl Schwarzschild, German physicist and astronomer, found a spherically symmetric solution to the equations. But there was a glitch. As smaller and smaller radii were considered, a radius was reached, now known as the Schwarzschild radius, where the equations revealed something called a singularity—mathematically the sort of thing you get if you divide by zero. Physicists don't like singularities. They usually imply regions of infinite energy or force. Really, most singularities tell us that something is wrong with our theory in the regions where they show their face. But *this* singularity was pointing to something new and downright awe inspiring about our spherical friends, stars.

When massive clouds of interstellar dust coalesce, condense, and begin to radiate, a star is born. Within a few billion years from the time of their birth, all stars age and eventually die. But they have a very interesting afterlife. After a life of burning, stars use up their fuel and cool, and with the lack of outward radiation pressure, they eventually collapse under their own inward gravitational pull. In 1931, Nobel Prize–winning Indian physicist Subrahmanyan Chandrasekhar showed that when all the mass of a dying star collapsed to within a small enough volume, it formed an enchanting object called a white dwarf—a quiet remnant of the former star with the pressure of its own constituent electrons holding it up against gravity. One day, our sun will become a white dwarf, shrinking roughly to the size of Earth. In 1939, Robert Oppenheimer and George Volkoff, with the work of Richard Tolman, showed that for stars more massive than the sun, even just one and a half times bigger, their gravity would be too great for their constituent electrons to hold them up. These stellar remnants collapse further until finally their neutrons take up the slack, pushing back against gravity. The result? Neutron stars. For stars greater yet, three times or more massive than the sun, even neutrons can't fight gravity. The nuclei collapse—and then our theories teeter on the edge of our understanding. In step black holes.

Black holes became a theoretical reality with the Schwarzschild solution of general relativity, and they became a physical possibility with the understanding of stellar evolution. In 1958 (around the same time that Leon Cooper found his solutions to superconductivity), one of my heroes in physics, David Finkelstein, discovered something truly remarkable to make the black hole story more interesting yet.

David Finkelstein is quiet, sage-like, and beaming with genius as if the entire cosmos were contained in his head. So inspiring is he that it's not altogether surprising that pioneers of the two main competing theories of quantum gravity, Lee Smolin and Lenny Susskind, were both mentored by David. In fact, I became such a fan of David that, in 2014, I hosted a symposium at Dartmouth to celebrate his lifetime achievement.

What David wanted to understand was how a beam of light moved in the warped space-time around a black hole. After all, it was the observation of the bending of light from a distant star around our sun that confirmed Einstein's idea that gravity was, in fact, the warping of space-time around a massive object. But, as David found out, the movement of light around a black hole was even more bizarre. By an ingenious reshuffling of the equations that govern space-time, David found that there was a spherical bubble-like region surrounding the singularity in Schwarzschild's solution such that if anything entered this region, including light itself, it could never escape. That's why John Wheeler coined the term *black hole* to describe these things, in fact. If no light could escape the Schwarzschild region surrounding the singularity, you'd never be able to see it. Anything entering this region would essentially disappear into blackness. What David had discovered was a one-way invisible spherical surface, which he called a horizon. It was a horizon no one could see beyond, not completely dissimilar to our visual horizon into the universe's past, making its study all the more intriguing.

When David made his calculations, black holes were still a subject of sci-fi novels, a playground for the imagination, but they were beginning to be understood as well. While some physicists, like Lee Smolin, speculated that black holes spawned baby universes at their singularities, we also learned that black holes can grow in mass by consuming matter and that they can radiate, due to quantum effects near the event horizon of the black hole. David's work made the study of black hole physics concrete. The event horizon was a definitive, albeit intangible, mathematical element to work with, one that might even shed light on the structure of our universe and the ancient cosmic horizon. To better understand how, we need to look at sound—specifically, how sound moves in water.

Canadian physicist Bill Unruh found this brilliant analogy in terms of sound that captures a great deal of the physics of black holes. Bill is one of Canada's and the world's most revered theoretical physicists. I spent a half year at his home institution, the University of British Columbia in Vancouver, to work on my PhD dissertation. Bill is a big man with a full beard and usually wears overalls. He has a tendency to intimidate other physicists and is quick to pounce on any inaccuracies that may present themselves, but he was always kind to me even when I said dumb things. His mastery of finding analogies for physics concepts spoke loud and clear to me one day at the University of British Columbia when he found a mistake in the first seminar I ever gave and proceeded to suggest a correction. A year later his proposal worked impeccably.

We can calculate the speed of sound in water using the basics of wave mechanics. Consider the following equation:

$$c^2 = \frac{K}{\rho}$$

This equation relates the speed of sound, c, to the stiffness of the medium, K, and the density of the medium, ρ. The equation is telling us that the speed of sound increases with the stiffness of the material but decreases with increasing density. Sound will travel slower in

FIGURE 11.1. Canadian theoretical physicist Bill Unruh. *Bill Unruh.*

a denser gas, such as oxygen, rather than helium, yet will travel faster through stiffer materials, like solids. Though solids are more dense than gases, and so one might think sound would travel slower in them, solids are far stiffer than gases, speeding up sound travel.

So to understand the black hole horizon, Bill imagined a fish going downstream while a fishy friend remains upstream. At some point, the downstream fish takes a plunge down a waterfall. The speed of the water in the waterfall far exceeds the speed of the water upstream because gravity has given it a boost. Rushing downward, the fish screams, hoping that his friend will hear, "Hey, I'm falling!" But sound is a wave, and as the above equation illustrates, it moves at a fixed velocity in a uniform medium. If the speed of the waterfall is much faster than the speed of the fish's sound wave moving upstream, then the sound wave

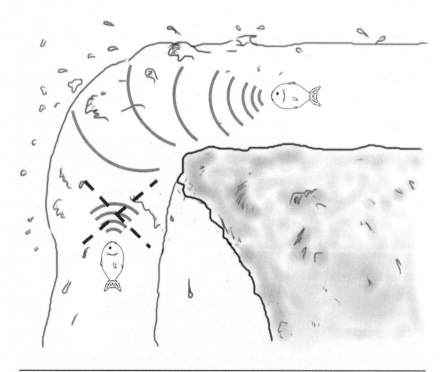

FIGURE 11.2. A sonic horizon can be understood with a waterfall. A fish emits a sound, denoted by the circles, but the speed of the sound is much slower than the speed of the waterfall, so it never gets to the upstream fish.

will never get to the other side of the waterfall for his friend to hear him. An uphill battle, lost. To the fish falling down the waterfall, the sound can be heard, but to his friend on the other side, there is silence. The edge of the waterfall is a sonic horizon. To the upstream fish, his friend simply disappeared—out of sight, out of earshot, and, for fish, out of mind. Of course, if *he* called out to his lost friend, the sound would travel nicely downstream and over the edge of the waterfall, aided by the flow of the water. This is how light behaves around the event horizon of a black hole. Light can enter the black hole easily, but exiting is another failed story.

The black hole solution in general relativity had a predictive power that few physicists anticipated—the surprising reality of the event horizon. According to the black hole solution, if the fish fell through the event horizon, no matter how hard the fish tried to communicate with his friend outside the black hole, his message would never escape to the other side of the horizon. Even more sad, once the fish falls through the

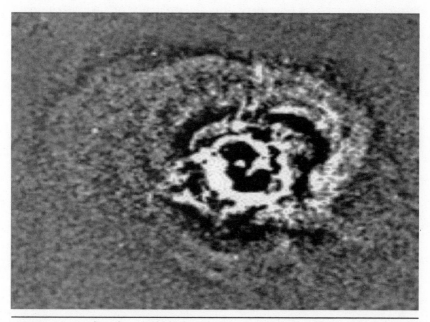

FIGURE 11.3. The white and black regions represent sound waves "played" by a black hole in the Perseus cluster.[1]

horizon of the black hole, he would have no hope of coming back out. Not even a salmon could thrash his way out of this.

Not only do black hole horizons have a sonic quality, but it was recently discovered that some black holes play a drone-like song. Figure 11.3 shows the sound wave generated by a black hole at the center of a galaxy in the Perseus cluster. The note of the black hole's sound was identified as a B-flat fifty octaves below middle C on a piano.

The existence of horizons is a general feature of Einstein's theory and has serious consequences for the discussion of the space-time structure of our universe. This is true of both black holes and the cosmic horizon. The cosmic horizon, however, is a bit different from an event horizon. Unlike a black hole, it is a two-way avenue, where light and matter cross both ways, depending on the interplay between the expansion of the universe and the passage of time.

Though black hole horizons are distinct because of the immense gravitational forces involved, they have helped us understand how a horizon can act as a boundary. It is the existence of such a boundary, the cosmic horizon, at the time that the CMB light was released, when the first stable atoms were formed, that created resonances in the CMB anisotropies. Just like bridges on a guitar provide the boundaries necessary for a string to resonate and create notes, the cosmic horizon allows for discrete notes in the matter perturbations of the universe. What causes these vibrations fixed by the cosmic horizon fret board? This is where quantum mechanics comes in.

12

THE HARMONY OF
COSMIC STRUCTURE

My search for the link between music and the structure of the universe led me to take a yearlong sabbatical at Princeton University in 2011 to work with David Spergel, a lead scientist of the Wilkinson Microwave Anisotropy Probe, a space telescope that made some of the most precise measurements of the anisotropies in the CMB. Right down the hall from my office was Jim Peebles's office: he was one of the first cosmologists to confirm my hopes that the correct way to view the CMB was not through the lens of the anisotropies but through sound oscillations.

Peebles and his graduate student Jer Yu were the among the first to validate Pythagoras's and Kepler's intuitions of a musical cosmos.[1] They discovered that the early universe generated sound waves with wavelengths extending 300,000 light years across—the size of the universe when the first stable atoms were formed and the CMB radiation was released. These sound waves contributed to the eventual creation of large-scale structures in the universe. Peebles and Yu summarized at the beginning of their groundbreaking 1970 paper, titled "Primeval Adiabatic Perturbation in an Expanding Universe," that "the possible discovery of radiation from the primeval fireball opens a promising lead toward a theory of the origin of galaxies."[2]

FIGURE 12.1. A snapshot of the universe 13.7 billion years ago revealing the light emitted from the moment that electrons combined with protons (recombination). *WMAP science team.*

The plasma of tightly coupled electrons, photons, and protons moved with the perfect synchronicity of square dancers and if left undisturbed would have remained still as the surface of the calmest ocean. But a primordial quantum disturbance due to quantum uncertainty caused the plasma to vibrate in such a way that regions of higher density transferred energy to regions of lower density, resulting in the propagation of sound waves.

This revolutionary realization was that in order for large-scale structure formation to evolve out of the early universe, it would have had to contain certain irregularities on the order of one part in ten thousand. In other words, if the average temperature was ten degrees then the irregularities would deviate from the average by one one-thousandth of a degree. In 1992, these long-sought-after anisotropies were finally discovered experimentally by the COBE satellite.

The black, white, and grey spots in the CMB map represent the fluctuations above and below the average, uniform energy (or temperature). These are the sound waves of the early universe, the music of the universe, and the first stages of structure formation. Light waves are produced by accelerating charged particles and do not require any medium to propagate. Sound waves, on the other hand, cannot exist without a space-time medium to push and prod; they don't like a vacuum. Sound waves are mechanical. They carry speech, music, and noise to our ears; they are the vibrations of the medium, whether they reach our ears or not. When there is an initial disturbance (like striking a drum), the vibrations cause neighboring particles to oscillate back and forth. These particles, in turn, cause their own neighbors to vibrate, leading to a series of compressions and rarefactions that propagate the wave through the medium. When these oscillations in air pressure reach our ears, our brain interprets them as sound.

The different fluctuations seem to have no interesting features, but by using the Fourier idea, we can decompose the CMB map into pure waves. Remarkably, we see in the curve in Figure 12.2 what is characteristic of sound waves. The X axis represents the frequency of the CMB sound wave, and the Y axis represents how loud the sound wave

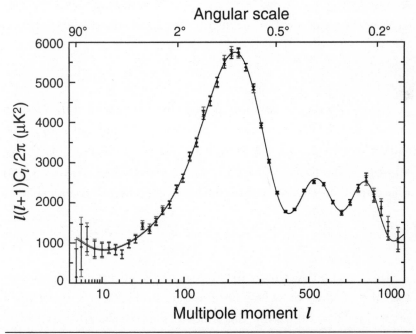

FIGURE 12.2. The Fourier transform of the CMB anisotropies, revealing the sound waves and resonant frequencies in the CMB oscillations. *WMAP science team.*

is. The peaks represent the resonant frequencies. As we will see later, these resonant frequencies play a dominant role in seeding the structures in the universe. Instruments make sounds, and we can understand much of the physics of the CMB by making an analogy with how musical instruments function.

For concreteness, let's consider how instruments generate sound. Blowing into the mouthpiece of a saxophone, for example, generates pressure waves carried by the air molecules inside the instrument. The reed vibrates at a wide range of frequencies and creates the source of the sound, and some initial energy is needed to establish a pressure difference in the air. There are many possible airwaves that can fit inside the horn. In the case of a guitar, its fundamental frequency, the lowest frequency it can produce, is the longest wave pattern that can fit between the end points of a guitar string. Another way of describing the fundamental frequency, or simply called the fundamental, is to imagine extending it such that one complete cycle of its wavelength is twice the length of the guitar string, bringing us to the following useful relationship between the length of the instrument, L, and its wavelength λ:

$$\lambda = 2L$$

There is another useful formula that relates the speed of the wave to its period, T, which is the time it takes a wave to undergo a full cycle. We can quickly deduce this formula by using Newtonian thinking. We know that the distance covered is simply the product of velocity and time elapsed. Therefore, we find the following powerful relationship:

$$\lambda = vT$$

This equation succinctly determines the wavelength of the fundamental tone if we know how fast the wave is moving and its period of oscillation. By multiplying this fundamental wavelength by an integer, we get the higher harmonics (or higher partials) that also play an

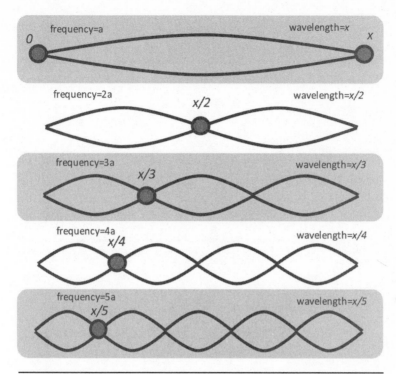

FIGURE 12.3. Standing waves on a string fixed at both ends. The nth harmonic represents a wavelength equal to 2/n of the length of the string. In this convention, the length 2L = x.

important role in the unique quality of an instrument's sound known as its timbre.

Timbre is a very important concept. Consider two instruments: a flute and a clarinet. They can play exactly the same note but retain their own distinctive sound, which enables you to differentiate them. This unique signature is what we call timbre. When an instrument generates a note, it does not generate one single frequency. Recall that a string can be seen as a sequence of masses attached to springs. The natural frequency is dictated by an infinite set of springs. When a string is plucked, it vibrates with a wide range of resonant frequencies on top of the fundamental. Depending on the material that the string is made of,

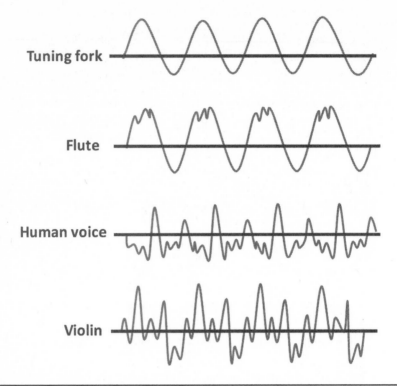

FIGURE 12.4. The physical properties of different instruments yield a unique signature produced by the characteristic damping of higher harmonics.

the amplitudes of specific higher harmonics will get dampened. They will die out because certain natural frequencies lose energy due to friction and consequently lose amplitude. What's left results in a unique signature of different amplitudes for the higher frequency harmonics, and thus a unique sound.

The timbre is a characteristic vibrational energy of a particular object transmitted through the air. Think of one mass passing on its characteristic motion to the next mass via a spring. But just as springs do not oscillate forever, due to damping effects such as friction or heat dissipation, an instrument's coupling efficiency to the air is imperfect and is frequency dependent. Higher frequencies on the piano couple more efficiently to the air than do lower frequencies. Using Fourier's idea, the sum total of the frequencies produced by an instrument is the sound spectrum for an individual note. Our auditory system picks out the fundamental as the "pitch" and interprets the higher order harmonics as the timbre of the instrument.[3] It is interesting to note that, though a tuning fork generates a perfect sine wave, the sound is not as musically rich as that of a violin, which has a richer spectrum of higher harmonics.

There is no fundamental law of physics that requires the early universe to sustain sound waves. Yet if we think of the early universe during its plasma epoch as an instrument, acoustics shed light on structure formation. What Peebles and Yu discovered is that the CMB was a medium where acoustic oscillations were initiated and sustained for 300,000 years. If this is the case, then we can apply all of the concepts just described to understand the consequences of the sound waves propagating through the CMB. Shortly after the big bang, energy imparted into the plasma from a previous epoch (most likely cosmic inflation) created sound waves.

Two forces worked to sustain these sound waves: gravity and radiation. Gravity caused matter to clump and set up an overdensity in pressure, which left unimpeded would simply have collapsed too quickly to form any interesting structures across the cosmos. Luckily for us, light exerts a restorative force just like a spring does, and the primordial plasma was full of photons of light. When a photon scatters off an electron, it undergoes a change in momentum and exerts a force according to Newton's second law. Many photons scattering off the electrons in the plasma exerted enough pressure to resist further gravitational collapse. As a result, the plasma wave expanded, and the pressure decreased. This is where gravity came back in to compress the plasma, and this harmonic dance is what became the first sounds in the universe. According to the standard model of cosmology, the cosmos hummed these sounds in harmonic sequence during the first 300,000 years following the big bang like a perfect cosmic instrument.

The particles in the CMB are highly interacting, and the speed of sound is very close to the speed of light. Armed with this knowledge, we can apply this amazingly simple formula for the wavelength of the fundamental frequency to the CMB plasma.

Given that the wave traveled close to the speed of light and that it traveled for 300,000 years, we can use the formula for a sound wave that was previously derived:

$$\lambda = \nu T = 3 \times 10^8 \text{m/s} * 3000,000Y \cong 1 \text{ Mpc}$$

We see that the distance that the fundamental sound wave in the CMB covered was on the order of a million parsecs! Amazingly, when we look at how the galaxy clusters that Geller and Huchra discovered are distributed, we see that they occupy islands precisely of this size. The tiny sound waves that Peebles discovered did develop over a period of billions of years into the large-scale structures we see today, and it all started with sound waves. But if sound is nothing more than a blend of oscillating standing waves, how could it develop into galaxies and stars?

Many musicologists believe that music is structured sound. So if the sound structure developed into cosmic structure, is the universe musical? At recombination, when electrons bond with protons to form hydrogen, gravity wins over, and the tones transmute into rhythms. These rhythms represent hydrogen gas collapsing into central regions to form the first stars and protogalaxies. We will discuss the process of star formation in more detail in the next chapter.

To summarize, we have a picture of the first moments when structure formed out of the unstructured early universe. The first patterns were a combination of sound waves of various frequencies. In an ideal resonator, like an idealized string, all frequencies would be generated at the same amplitude or loudness. But when we analyze the CMB data in terms of its component frequencies, we see that there is a loudest peak (called the acoustic peak). This is similar to what happens in musical instruments, when the acoustic peak often corresponds to the note that one hears. The other peaks reveal the timbre of the sound and depend on other physical factors, such as the material of the instrument. Likewise, we can expect the other acoustic peaks in the CMB to yield

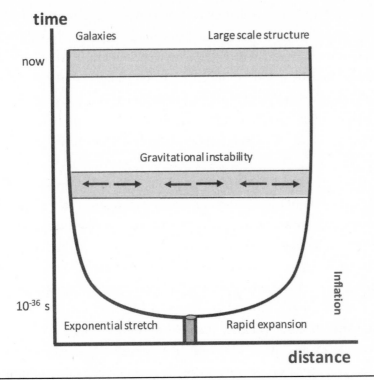

FIGURE 12.5. Diagram of a ten-billion-year time span focusing on the epoch of gravitational instability after recombination until large-scale structure formation in the present era.

information about the physical makeup of the universe. And marvelously, they do! For instance, they tell us about the dark matter that must have existed in the early universe.

Going back to what I learned from Leon Cooper, when an analogy breaks down, the possibility to discover something new emerges. For one, during the time that the cosmic plasma played its music, the universe had been expanding from the big bang until the first light elements formed. This means that the waves that were generated early on got stretched with the expansion. To make sense of this, imagine drawing a line on the surface of a balloon. As the balloon gets filled with air and expands, the line itself expands with it. Similarly, light waves will get stretched as space expands.

Second, a simple calculation reveals that the sound waves travel close to the speed of light. And what does the CMB actually sound like? Some cosmologists have turned the frequencies of the CMB into sound, and though it is not very musical, it is not pure noise either. What is fascinating is there was an original quantum sound, which caused the first primordial vibrations in the plasma, and though this sound is categorized as white noise, its beauty is in the eyes of the beholder.

If the structures in the universe started out as sound waves, do they lead to a more complicated structure that is also musical in nature? Does our musical universe have analogues of tone, melody, harmony, and rhythm? I think so.

One hundred fifty million years after the big bang, sound waves and hydrogen developed into stars, and the stars clustered into galaxies, but it's not that straightforward. As pressure density waves grow in amplitude with clustering matter, the initially simple sound equations become highly nonlinear. Moreover, the attractive gravitational energy in the primordial sound waves, which are comprised solely of ordinary baryonic matter, are too tiny to collapse into the network of galaxies we see today. Some missing form of matter is needed to amplify the gravitational attraction to collapse into stars and galaxies—dark matter. Dark matter cannot directly interact with light and visible matter. By watching how fast stars rotate within galaxies, cosmologists have gathered observational proof for the existence of dark matter. Dark matter provided the timbre in the harmony of the growing universe.

According to Newtonian mechanics, the velocity of a star rotating around a massive galaxy will decrease the farther away it is from the galaxy's center. But Vera Rubin discovered that the velocities of the stars did not decrease and, instead, approached a constant. The dark matter, baryonic matter, and photons are all quantum fields, and their associated particles were created from the vacuum in the earliest stages of the universe. One of the main research avenues in cosmology is to understand the correct physics responsible for the creation of the elements in the universe's primordial plasma.

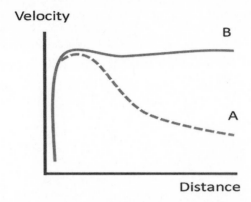

FIGURE 12.6. A typical rotation curve of velocity of a star around a galaxy. The dashed line represents the prediction from Newtonian gravity. The solid line is the observed velocity, requiring some missing dark (nonradiative) matter.

Equipped with the right amount of dark matter, computer simulations reveal the formation of large regions of filamentary web-like networks of dark matter and hydrogen gas. At the nodes of these filaments, hydrogen gas coalesces, similar to water droplets on a spiderweb after a rainfall. It is in these regions, called protogalaxies, where hydrogen gas gets gravitationally concentrated into the first stars. Through nuclear fusion, the immense gravitational pressure in the star converts hydrogen into heavier elements. These first-generation stars can be up to one million times more massive than our sun. The first generation of stars lives on the order of one hundred million years and eventually dies through supernova explosions.

Over the course of cosmic history, there have been three generations of stars, with important differences among them. The earliest stars, Population III, are made up of hydrogen and helium. The second-generation stars, Population II, are metal poor and smaller than Population III. Contemporary Population I stars, like our sun, are metal rich and much smaller. The first-generation stars were forming roughly 250 million years after the big bang and only lasted a few million years. There may be around ten billion trillion stars in our observable universe.

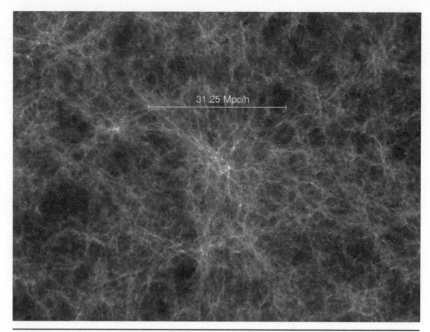

FIGURE 12.7. The filamentary structures in the large-scale structures of stars clustered in protogalaxy nodes of the intersecting filaments.

I was giving a lecture at the University College of the Cayman Islands in the winter of 2015 on cosmology and music. I usually use my saxophone to demonstrate some of the very ideas that I have discussed so far in this book. Attending the conference was astrophysicist Ed Guinan, who is known for being the codiscoverer of the rings of Neptune, which turned out to be two moons. Over a glass of the local Caybrew beer, Ed told me that I should look into the field of helioseismology, which concerns the study of sound waves on the surface of stars.

The sun is a nearly perfect ball with a surface of hot plasma. Turbulence creates sound waves on the sun's surface, similar to the wave patterns generated on a struck bell. I had a big smile on my face after learning that all of the stars in our universe play a tone. The initial sound generated by the primordial plasma became stars, which in turn create sounds. I discovered that a few astronomers started to use helioseismology as a means to study the sound waves on stars to learn about the inner structures of the star. My vision of a musical universe was more than an analogy; I realized it was becoming literal.

Waveforms from the early universe formed stars. Stars, in their tumultuous fusion of elements, produce sounds like tones. They organize themselves into larger structures, such as binary systems or clusters— the equivalent of "musical" phrases. What's more, the millions of stars within galaxies organize themselves into self-similar, fractal structures, like the fractal structure found in Bach's and Ligetti's compositions. I was amazed at the degree to which the organization of cosmic structure mimicked music structure. When an analogy goes beyond your expectations, you can't help but wonder if the analogy is the truth.

I end this chapter with a quote from the revolutionary composer John Cage:

> Definitions: Structure in music is its divisibility into successive parts from phrases to long sections. Form is content, the continuity. Method is the means of controlling the continuity from note to note. The material of music is sound and silence. Integrating these is composing.[4]

13

A JOURNEY INTO
MARK TURNER'S QUANTUM BRAIN

One night at the Village Vanguard, during the intermission, I could not believe the words I heard from one of New York's finest jazz sax players. He said, "When I'm in the middle of a solo, whenever I am most certain of the next note I have to play, the more possibilities open up for the notes that follow." One of my living heroes on the tenor sax, Mark Turner, was saying this. It was the spring of 2002, and after years of searching for a deeper relationship between jazz improvisation and physics, Turner's words confirmed that I was not hallucinating. His affirmation trumped all the snubs from other musicians and scientists whenever I would engage them with scientific concepts and its relation to music. Turner's insight about the potentialities that open up in the middle of an improvisation relate directly to the quantum mechanical uncertainties in the early universe. His statement gave me an insight into the question: How do all the matter and fields and its associated cosmic structure arise from a state of emptiness? After all, some sort of magic must have happened in our featureless early universe to ignite the first structures.

Mark Turner had an interesting trajectory for a jazz musician. He started playing the clarinet in elementary school. In college, he had a short love affair with commercial art. He later found his true love, the tenor sax, and went to the prestigious Berklee School of Music, eventually moving to New York. For a number of years, Turner worked at Tower Records in Manhattan before working full-time as a jazz player, mostly as a sideman. All the while, he worked on his music. And in the end, Turner, in the opinion of many horn players, managed to emerge as a living incarnation of John Coltrane.

Many jazz musicians suffer from a Coltrane inferiority complex. Coltrane possessed both natural musical talent and practiced harder than what most of us think is humanly possible: Like Charlie Parker, he would often practice up to fourteen hours a day! But one of the qualities that canonized Coltrane was his stylistic versatility as he left few musical stones unturned. After releasing his signature hard-bop masterpiece that established his harmonic mastery, *Giant Steps*, Coltrane explored both the microtonal system of Indian music and various polyrhythmic genres of Africa. And he didn't stop there. Near the end of his life, he mapped out the cosmic sounds of free jazz, exemplified by the piece "A Love Supreme." He also came up with his own, harmonically lush sheets of sound: a rapid fire of arpeggios,[1] which creates a perception of a chord. Naturally, any contemporary tenor sax player striving to make a mark will live in Coltrane's shadow.

Turner is one of the few tenor players since Coltrane who has been able to create his own style. He did this through the practice of transcription, enabling him to analytically dissect and amalgamate the works of several greats, primarily John Coltrane, Joe Henderson, and Dexter Gordon.[2] This is not cheating: it is completely respectable and essential for musicians to devote themselves to mastering the works of the greats, and most do—including myself. But what really gave Turner his unique voice was his exploration of Warne Marsh—a horn player known by aficionados but less familiar to the public. Turner's breakthrough arrived when he managed to combine the "loopy" style of Marsh with Coltrane's sheets of sound.[3]

Marsh was a student of composer and pianist Lennie Tristano. Born in Chicago, Illinois, in 1919, Tristano became fully blind at age six. Nonetheless, he attended the prestigious American conservatory of music in Chicago, where his aunt took notes for him. After moving to New York, Tristano developed a highly harmonic and improvisational approach to bebop jazz, which was summarized in a statement he made in an interview: "I don't compose anything . . . you see that's the difference between other forms of music and Jazz. The music is already in your head and what you do is let your hands reproduce what you hear as you hear it. So what you come up with is something completely spontaneous."[4]

Let's not be fooled. Doing this is not as simple as it sounds, especially when we think about the meaning of the word *spontaneous* in the context of improvisation. Going back to my first encounters with free jazz, I naïvely interpreted spontaneous playing as random playing: "Just play whatever comes to mind—press some keys on the sax and blow." Superficially, it may seem that's how these guys play, but this type of spontaneity is a function of years of practice, memorization, and making mistakes choosing "wrong" notes. The magic of improvisation is captured in Tristano's statement: "The music is already in your head." But how does a successful improviser get the music in his or her head? To accomplish this feat, Tristano's method required a deep understanding of music theory and harmony and then embodying the theoretical understanding: he would have his students memorize and sing complete solos from past jazz masters. This repertoire would train the inner ear of the musician and enable him or her to create more meaningful music spontaneously.

With Turner's statement in mind, I wondered if there was a science behind improvisation. This may have been an attempt to rationalize improvisation, but in fact, during 2002, I learned that it is less about music being scientific and more about the universe being musical. Music along with improvisation, which had inspired me to play the horn in my teens, had been there all along to help me understand the inner working of quantum mechanics—and by extension the structure of the universe. But I needed a catalyst—Mark Turner—to see it.

Let me quote Turner again: "When I'm in the middle of a solo, whenever I am most certain of the next note I have to play, the more possibilities open up for the notes that follow." The converse is that the more confused he was about the next note, the fewer possibilities opened up for him. At the moment that I heard this, I realized I would've benefited from hearing this a lot earlier. I had a big smile on my face and thanked him. Even today he has no idea how important that discussion was to me nor how his statement shook me free of my confusion about one of the most sacred principles in quantum mechanics, necessary to truly understand how the universe could have used this quantum magic to create the stars, galaxies, and us: the Heisenberg uncertainty principle. It's time I told him.

The world is pervaded with uncertainty, but the idealistic world of classical macroscopic physics isn't. According to the physical laws and equations governing the macroscopic world, such as electromagnetism and Newtonian mechanics, we can in principle know the future behavior of objects no matter how complex or how many interacting particles there are. The great French mathematician Pierre-Simon Laplace spelled out this philosophy: according to his analysis, once the initial position and velocity of objects in a physical system are specified then their trajectories can be known with certainty in the future. I say "in principle" because once we move beyond the classical realm into the quantum realm, uncertainty plays a fundamental role.

Quantum mechanics was born out of efforts to account for a handful of findings that were initially thought to be minor experimental anomalies. One was Ernest Rutherford's series of experiments with gold foil, which established that atoms were mostly empty space, with a massive, positively charged nucleus surrounded by a diffuse cloud of negative charge. That cloud was thought to be filled with orbiting electrons, which posed a problem. In the macroscopic world, orbiting objects necessarily accelerate toward the center of a circle. But when a charged particle like an electron accelerates, it loses energy by releasing radiation in the form of electromagnetic waves. This loss of energy would cause the electron to rapidly spiral inward, not allowing for stable atoms. Without stable atoms, there would be no stable molecules and no chance for life—bad news. And yet both atoms and molecules are clearly stable, worse news for conventional physics.

To make matters even more difficult for classical physics, other experiments were showing that when one shines a continuous array of light into a gas made up of hydrogen atoms, only a discrete set of light colors emerge out of the gas. It's like pressing all the keys on an organ, expecting to hear a texture of sound but instead having only two notes come out. The electromagnetic theory was unable to explain this experiment. Somehow the continuous cloud of orbiting electrons of the atom had to be replaced by something "note-like."

Solving these two riddles took several discoveries. First, Max Planck and Albert Einstein showed that light, thought to be a purely wave-like phenomenon, could also behave like particles. Light, they proposed, could come in a bundle or packet of energy called a photon and could, like a billiard ball, knock a bonded electron out of a metal. But there was a major twist. At the time, physicists thought a beam of light was like water coming out of a hose. If the volume of water is increased, the water will have more momentum. The same behavior was expected of light waves. But something different was seen in the photoelectric effect: no matter how intense the light, the same number of electrons came flying out. However, increasing the frequency of the light—essentially, making it bluer—caused the light to hit more electron homeruns. Two conclusions were drawn from this experiment:

1. Depending on the situation, light can behave like either a wave or a particle.
2. The kinetic energy that a beam of light imparts to electrons is related to the frequency of the light and not the intensity.

These were incredible conclusions. Many physicists at the time thought that the photon was still intrinsically a wave, and its particle-like behavior only emerged when the waves bundled up—much like Eno used the Fourier idea to add up waves to make blip-like sounds. But this explanation was too naïve. Even Einstein stated, "Nowadays every Tom, Dick and Harry thinks he knows [what the photon is], but he's wrong."[5]

I still remain amazed at Einstein's genius for getting to the heart of the matter. From these two observations, Einstein discovered an elegant and fundamental relation for the energy of light that nailed down the photoelectric effect.

$$E = hf$$

The equation relates the energy of a photon, E, to its frequency, f, and reflects that the photons are discrete bundles and not continuous, as we normally envision light to be, and, finally, to h, the famous Planck constant that he inherited from physicist Max Planck. Planck had studied the radiation emitted from objects in thermal equilibrium with their surroundings, so-called black bodies, and determined that light had to be quantized according to the relation $E = hf$ in order to explain the observations. Einstein applied this equation to reflect the discrete energies that light in the photoelectric effect can carry, and his relation successfully explained the results.

It took a flash of genius by the young doctoral student and violinist Louis de Broglie to use the work on the photoelectric effect to solve the problem of the orbiting electron. De Broglie asserted that if Einstein tells us that waves can have particle behavior, why can't particles have wave-like behavior? The key to de Broglie's solution was to associate a particle-like property, momentum, with waves and to imagine electrons not as miniplanets orbiting a nucleus but as standing waves on a string.

As we discussed, a string can undergo oscillatory vibrations or periodic waves. Fixed at both endpoints, the string will resonate at certain frequencies when plucked. The resonance comes from the fact that waves on the string travel in both directions and either add or cancel according to the Fourier idea. Standing waves, whose nodes stay fixed while the crest moves up and down periodically, are created. De Broglie hypothesized that the momentum of the electron could be associated with the wavelength of the standing wave orbit, similar to the relationship between a photon's energy and frequency, which he described mathematically as:

$$\lambda = \frac{h}{p}$$

In the above equation, p is the momentum of the electron as it moves around the center of the nucleus and is the wavelength. It is amazing that this equation is a physical reality, for it states that an electron's "orbital" wavelength, a wave-like property, is related to how fast it is going around the nucleus, its momentum. The larger the wavelength, the slower and lighter the particle—recall that momentum is the product of the mass and velocity of a particle.

De Broglie's equation holds for any form of quantum matter, not just electrons. Planck's constant sets the scale for the wave-like nature of particles. It is a tiny number, which means that we don't see the waviness of macroscopic matter because we are moving slowly compared to the fast quantum particles zipping around the atom. If we were very tiny, we would see our inner waviness. De Broglie's connection between the wavelength of a particle and its momentum is at the heart of the famous uncertainty principle. And it was Werner Heisenberg that was able to precisely formulate it.

A good way to understand the uncertainty principle is to consider a wave where we have complete certainty in its frequency, like a pure tone. Now, I ask you, where is the wave? The wave with its many periodic oscillations is distributed across a very large distance, meaning that a wave of definite frequency will have an arbitrary position. Now let's consider a traveling wave pulse, which only exists for a short duration of time, like a beat. I can localize where the pulse is, but its frequency is not well defined because a frequency requires many repeating cycles, and a pulse does not have enough width to define a definite frequency. This is Heisenberg's uncertainty principle: it says that the more you can know about the position, the less you can know about its frequency, and vice versa. But we just learned that the frequency is proportional to the momentum, so the more one knows about the momentum of a particle, the less one knows about its position, and vice versa. Stated mathematically:

$$\Delta x \simeq \frac{1}{\Delta p}$$

Where Δx is the uncertainty in position and Δp is the uncertainty in momentum.

This is incredibly profound. When scientists want to understand nature, they use instruments to probe and measure it. What the uncertainty principle tells us is that no matter how careful we are, no matter how precise our instrumentation, we can never pin down both the particle-like and the wave-like properties of a quantum entity, whether it be a photon or an electron, a quark or a neutrino. The uncertainty principle is a statement that is fundamental to nature, to the universe, whether we are there to measure it or not.

So what was it about Mark Turner's improvisational insight that changed the way I thought about the uncertainty principle? He said that the more certain he was about what note he was about to play, the more possibilities opened up for him for the notes that followed. Let's rephrase the uncertainty principle accordingly: the more certain the momentum of a particle, the more possibilities exist for it to be in a wide range of positions. This gets to the heart of quantum mechanics; our uncertainty is simply a reflection of a quantum particle having less restriction for a particular physical attribute.

What the uncertainty principle really reflects is the fact that a quantum entity is neither a wave nor a particle but contains both attributes at the same time. The essence of this principle is the Fourier idea. One can create a wave pulse from adding a handful of pure waves of definite frequencies. Similarly, the particle-like property (a pulse) can emerge from the wave-like property and vice versa.

In quantum mechanics, the Pythagorean theory of the harmony of the spheres was finally realized but on a microscopic, not macroscopic, level. In de Broglie's hypothesis, every "orbit" of the electron is a wave corresponding to a pure tone. Matter and waves are one and the same. Neils Bohr called this idea—that quantum matter can have both wave-like and particle-like properties—complementarity. But to truly understand the origin of this wave-particle complementarity, it took Erwin Schrödinger to write down perhaps the most important equation in physics and perhaps all of science—the Schrödinger equation.

Although the version of quantum mechanics that we just described—the uncertainty principle—has most of the features to help us understand the emergence of structure in the universe, we have to make it compatible with Einstein's theory of relativity, since this is the framework that gives rise to the expanding universe. When we do this, quantum mechanics will have two features that will be essential for understanding the structure of the universe: the vacuum and antiparticles.

After the discovery of quantum mechanics and Einstein's theory of relativity in the late 1920s, Paul Dirac wanted to understand the quantum theory of electrons moving around the atom near the speed of light. Dirac turned to relativity because ordinary quantum mechanics was only valid for nonrelativistic, Newtonian mechanics and failed in the relativistic realm. Generally, subatomic particles don't reach speeds near the speed of light, but the early universe was incredibly energetic. Particles moved like they were on steroids, quantum steroids.

When the electron moves at speeds close to that of light, there exist frames of reference where the electrons seem to have negative energies. Negative energy particles are troublesome to deal with, and physicists usually deem them as unphysical. However, Dirac did not take this traditional view. With a stroke of genius, he identified negative energy electrons as new particles, with a positive charge and positive energy. He proposed the first antiparticle. A year later, the positron, the electron's antiparticle as predicted by Dirac, was experimentally confirmed, winning him the Nobel Prize. The unification between relativity and quantum mechanics associated every particle with an antiparticle. From this, a concrete theory of the vacuum was born.

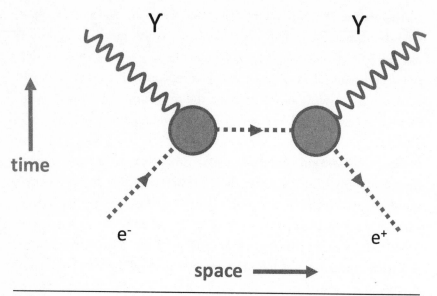

FIGURE 13.1. A Feynman diagram of an electron (e-) and an antielectron or positron (e+) annihilating each other to produce a photon of light (γ). The dashed lines represent the motion of the electron and positron. The wavy lines represent the motion of light.

If an electron and a positron collide, the total charge is zero. They would annihilate each other, and the energy from their masses would produce photons.

The reverse reaction would be if two photons, with an energy that's twice the mass of an electron, collide. The photons would create an electron and positron out of the vacuum.

It turns out this kind of thing happens all the time, spontaneously. That's because—despite our intuitions about the vacuum—the vacuum itself, on the tiniest of scales, is not empty, thanks to the Heisenberg uncertainty principle. Just as position and momentum are intimately connected via the uncertainty principle, there is an equivalent formulation with respect to time and energy.

$$\Delta E \simeq \frac{1}{\Delta t}$$

The time energy uncertainty says that the smaller the time interval a quantum process occurs, the wider the range of energies the quantum system can access and vice versa. When macroscopic beings like us look at empty space, the time scale is much too large for us to experience the uncertainty in energy; we perceive nothing, just empty space-time. However, if our eyes had the ability to probe shorter time scales, like a quantum camera with a high-speed shutter, there will, in fact, be a constant emergence of particles and antiparticles and their corresponding collisions. This feature of quantum field theory is tantamount to understanding the origin of the first sounds that led to the tones of the primordial plasma. In the early universe, we are dealing with time scales unfathomably tiny. The uncertainty principle tells us that the energy of the universe would be correspondingly highly uncertain, fluctuating constantly. These fluctuations disturbing space-time at the beginning of the universe would have been a chaos of emerging and colliding particles—exactly the kind of situation that would have left the anisotropies in the cosmic microwave background for twentieth-century physicists to discover. The infant cosmos wasn't uniform because the uncertainty principle doesn't allow it to be so at such high energies and short time scales. The uniformity and symmetry of the Copernican cosmos was broken by quantum physics at relativistic scales.

14

FEYNMAN'S JAZZ PATTERN

——Original Message——

From: <donharmusi >

To: <salexand>

Sent: Fri, Jun 1, 2012 8:42 am

Subject: Re: Jazz Saxophone

Dear Dr. Alexander, My name is Donald Harrison, and I also play the saxophone. I recorded a song called "Quantum Leap," which has led me to you today. I attached the song and some information regarding what critics, musicians, and I think of this jazz concept. Let me state that although I am not even in the ballpark in terms of your understanding of this topic, I would love to hear your thoughts on this music. I am approaching this concept more from my heart because I am limited in terms of expertise. In spite of my limitations, I hope you have a moment to listen to my song. Hopefully you will also have a moment to send a comment to me. All the best to you.

Thank you,

Donald Harrison

It seemed a typical day at Haverford College, where I held a faculty position, but on this particular one I received this e-mail from the legendary alto saxophonist Donald Harrison.

Donald and I had some long discussions, and his insights helped me push the analogy between quantum mechanics and jazz improvisation even further than I had done with Turner. In this case, the emergence of the universe's structure requires an understanding of the structure of the vacuum and the quantum motion of particles and fields in space-time. And the amusing surprise is that quantum motion according to Feynman's discovery looks very much like a jazz solo.

Imagine getting on the bandstand with your horn of choice in hand, let's say the trumpet. The ride of the drums is swinging in unison with the upright bass walking around a blues line. It's Miles Davis's "All Blues." The saxophone player has just finished his solo, and now it's your turn. No time to think—just play! You will have checked what notes are going to be agreeable with the harmony—the signature the song is being played in. Your ear will echo the meter and rhythm—the beats that are repeated per bar. If it's the blues, the signature will be distinct, recognizable, so that if at any given point you play any of the notes of the blues scale, you will, at the very least, sound harmonically agreeable with the rest of the band.

A more experienced improviser who has committed all the notes of the scale to memory will understand the relative harmonic importance between the notes. For example, the blues scale is comprised of only six out of the twelve notes in the Western sale: for the A blues scale, those notes are A, C, D, E, E-flat, and G. An inexperienced improviser could randomly play any of these notes and sound OK, but an experienced improviser would have developed a "vocabulary" with these notes. Wynton Marsalis says it best:

> In Jazz, improvisation isn't a matter of just making any ol' thing up. Jazz, like any language, has its own grammar and vocabulary. There's no right or wrong, just some choices that are better than others.[1]

Jazz vocabulary is analogous to phrases in spoken language. We use letters to make up words, and then string together words into phrases or sentences. Notes are like letters, scales and chords are like words, and jazz "licks" or patterns are like spoken phrases. So though simply playing up and down a scale or chord may not sound bad during a solo, an experienced jazz musician has committed many licks from the tradition to memory, which he or she can then deploy with ease during a solo. In a similar way, we can learn to write well by emulating language greats, such as a Shakespeare or a Toni Morrison. After all, why reinvent the wheel? I was once given a long riff by tenor saxophonist Eric Alexander, which had been passed on to him from George Coleman, and was told to get it under my fingers in all twelve keys. Though improvisation is about novelty, it is also about how much one has internalized phrases from the masters of the past—what I call phraseology. But what I've described here is simply a strategy of improvisation: fundamentally, improvisation is a lot more.

There are many strategies for improvising a coherent jazz line. Carefully applied, even repetition works. Hal Crook's book on improvisation, *Ready, Aim, Improvise!*, is a gem. But the main point is that jazz improvisation is *not a random process*. Improvisation is a function of memory, creativity, and, for mortals like myself, the number of hours you commit to practice.

One of the greatest improvisers of all time is Sonny Rollins. And one of his techniques is what critic Gunther Schuller calls thematic improvisation.[2] This was discussed in a historic article where Schuller analyzes a famous solo of a blues song called "Blue 7." Rollins begins

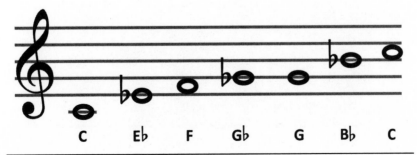

FIGURE 14.1. The notes of the C blues scale.

his solo with a three-note theme, which he uses as the framework to develop a more complex solo. As the solo continues, Rollins transforms the simple theme into intricate variations on the rhythm and harmony. The theme was like a skeleton guiding the evolution of Rollins's solo. Another important thematic improvisational strategy is simply to embellish the melody of the song during the course of the solo. It is common to get lost in the middle of a solo, and simply returning back to the melody or playing around the melody will get you right back in the pocket.

I was blessed to have some extended conversations with Sonny Rollins in the winter of 2015. When I asked him about thematic improvisation, he said, "I appreciate Gunther Schuller's article, Stephon, but I practice hard, and when I play, I don't play what I practice. You can't think and play at the same time. When I play, I don't want to play the music; I want the music to play me."

Now, imagine you are in the middle of your solo. In quantum mechanics, the act of observation actually disturbs the system: if an electron is not being watched, it will traverse many paths at the same

FIGURE 14.2. Sonny Rollins. *John Abbott.*

time. In the state of pure improvisation, according to discussions with Sonny Rollins and Donald Harrison, and from my own personal experience, there are moments when the player is not "observing" the notes being played, and like that quantum electron, the notes seem to do a quantum dance. If you play nothing, the groove goes on, just like the inevitable flow of time that occurs even when you're sitting still, doing nothing. Every improvisation is a new experience—not a reiteration of something past but of something never done before. A bass line may be familiar, but beyond that, it's a fresh palette. As most do when approaching something new, you proceed with caution. So, you restrict yourself to the seven notes of the blues scale. First, you play them slowly, but you quickly realize that you don't sound too bad. Your confidence goes up a notch. The band members, like any good jazz cats, are supportive and don't judge you—they give you space to just be with the notes. You have fun.

On the next gig, you will come back with that blues scale completely memorized and even some patterns that you memorized from a favorite Charlie Parker solo. You find some licks in his solo that you like and commit it to memory in all twelve keys. You realize that at any given time in your solo, you are aware of all seven notes at the same time. What's more, this familiarity means that you are intensely aware of the fact that the next note you play depends on the previous notes you played. The likelihood for playing one of these seven notes is conditioned by memory and repertoire, and this is happening in real time. Mark Turner's statement and insight becomes a reality.

This improvisational expression is at the heart of Richard Feynman's formulation of quantum mechanics with Feynman diagrams. A particle in classical Newtonian physics starts at some initial time and moves through space ending up at rest at a later time, tracing out a deterministic, continuous one-dimensional trajectory. Feynman realized that when a quantum particle moves between two points, *all* paths between those two points are considered, all paths are a quantum mechanical

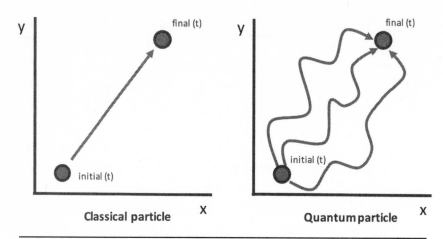

FIGURE 14.3. A classical path traverses a unique curve. To the right is a quantum path, which considers all possible paths between two points.[3]

possibility, even if they are not all equally likely—much as a musician considers all the notes in a scale before deciding which note to play in an improvised solo. Replace the notes with quantum particles and improvisation with probability, and the analogy is set.

When I came up with this analogy between jazz improvisation and the Feynman path integral, I thought I was going bonkers. So when Donald Harrison independently e-mailed me with a similar idea, it was comforting to find someone else who was equally bonkers. From the jazz side, Harrison was conveying the key idea of knowing only the beginning and ending notes, with nothing in between except time. The musician then improvises a musical path, connecting those two notes. The ending or "target" note is central to how the improviser traverses that path. In "Playing in the Yard," Sonny Rollins's solo starts with D and ends with the target note G: the two notes are harmonically related by a perfect fifth. The other notes in the scale trace out a path through time, connecting the originating and ending notes.

The experienced improviser would subconsciously consider all possible notes, or paths, the moment before actually playing a line. A path of notes is being played, but it is an integration of all the possibilities.

FIGURE 14.4. A path traversed by Sonny Rollins during an improvised solo.

How does the universe "consider" all these paths? Each path has a particular probability relative to the others. It is only when all the probabilities are added up for all paths that one gets the actual path that is most likely to be traversed—which is the one that is actually observed. Likewise, the experienced improviser will "integrate" the relative likelihood for each note.

In quantum reality, the path that results from adding up all these probabilities will be vague, representing the uncertainty inherent in a quantum system, which Heisenberg identified. How is this seemingly magical behavior possible? The trick is that each potential path corresponds to a quantum wave, and these waves have a unique property that particles don't: each wave can constructively or destructively interfer with other quantum waves—the good old Fourier idea at work! Most of the paths far away from the actual path will destructively interfere with other paths and are ultimately not seen. Likewise, other paths would constructively enhance each other, resulting in an increase in the probability that the observed classical path will manifest. This furthers the analogy and brings to light a fascinating question. Because notes are also waves, could it be that a kind of wave interference is occurring in the brains of "quantum" jazz improvisers like Coltrane, Rollins, Turner, and Harrison, as they decide which notes to play out of all the possibilities? Only an elaborate brain scan will tell. As it turns out, my colleague Michael Casey is studying brain scans of musicians and nonmusicians as they think a musical thought.

The Feynman path integral is an important part of conceptualizing quantum mechanics, making it easier for physicists to visualize and mathematically organize where a particle is going, just as some jazz practitioners have developed ways of conceptualizing the line that they're about to play. Feynman's path integral was the key conceptual leap that allowed physicists to understand how the intrinsic waviness and particle-like properties of quantum matter can work together to explain quantum motion. At high energies, Feynman and his collaborators discovered that particles and their associated waves had to be replaced with fields. The path integral also became the framework for quantum fields, which meant that not only the motion of quantum particles could be described but also their creation and destruction from a state of emptiness. Understanding how the improvisational nature of quantum fields function in a vacuum is essential to generating the building blocks of matter in the universe, which gave rise to the plasma that comprises the cosmic sea of photons, electrons, and protons in the CMB. What we will see is yet another musical aspect in the structure of the universe.

15

COSMIC RESONANCE

As we delve deeper into the sonic nature of our universe and the origin of its structure, we will see that much of the stuff in our universe emerged from the resonance of quantum fields, much like notes arise from the resonance of vibrating strings. Quantum field theory is currently the fundamental unifying paradigm of our universe. An example of a field familiar to most is the magnetic field. Unlike ordinary contact forces, magnets exert a force without touching each other because an unseen magnetic field emanates from both magnets. The field lines are invisible to the eyes but can be made visible by putting iron filings on a piece of paper around a bar magnet. The filings will bend from one magnetic pole to the other, indicating the field lines; the more the lines curve, the stronger the magnetic force. One of the mysteries in cosmology is that magnetic fields even exist across galactic distances, and we still don't know how and why.[1]

A closer inspection will reveal that the magnetic field lines are concentrated around the north and south poles of the magnet, suggesting that they fall away from the poles. So a magnetic field can be mathematically characterized by a function that has a direction at every point in space (for the curviness) and a number (for its strength). This type of function is called a vector field, and for a magnetic field, we put an arrow on top to signify its direction. Unlike a particle, which is described

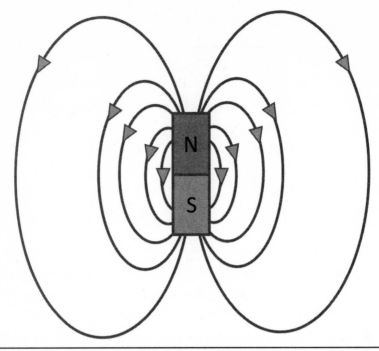

FIGURE 15.1. The lines represent the direction of the magnetic vector field from a bar magnet. The strength of the magnetic field, similar to gravity, falls off with the square of the distance.

as a single point, fields are entities that can be smoothly distributed throughout space.

When quantum mechanics was discovered, physicists were already aware of continuous fields such as electricity and magnetism. In fact, James Clerk Maxwell was able to combine the electric and magnetic fields into one entity, another vector field that is often referred to as a vector potential. In order for electrons to make quantum jumps across energy levels in atoms, they have to interact with and emit photons. And the physics for quantizing any field—including the vector potential of electromagnetism, the photon—is similar to how we obtained the simple sinusoidal vibrations of the string in Chapter 8. The same way we used the Fourier idea to express any complicated string vibration, we can do the same for the vector potential field and express it as an infinite sum of waves that differs by an integer frequency. For example the photon of a given frequency would be a pure sinusoidal wave of the vector potential with a quantized frequency. The photon field (quantized vector potential field) would therefore be an infinite collection of photons of integer values. Just as quantum spins in the Ising model can interact with each other to control the amount of magnetism, different quantum fields can interact with each other to give the plethora of behavior that we observe in our world.

By probing the structure of matter with particle accelerators, we now know that all matter and the carriers of the four forces arise from fields. As far as the visible matter in the universe is concerned, there are two basic types of fields—fermions and bosons. Bosons are the fields that carry the forces, and fermions make up the content of matter. In the case of the atom, the fermion is the electron, and the boson is the photon. They interact to cause the electron to emit a photon as it makes a quantum leap downward or to absorb a photon as it makes a quantum leap upward. Now, beyond the electron and the photon there are a handful of fermions and bosons. The quanta, or harmonic vibrations of the fields, are the particles. Three of the boson fields are responsible for the gravitational, electroweak, and strong forces. What's even more breathtaking is that all electrons—those in your body, those in stars, and those distributed across the universe—arise as vibrations from one universal electron field that permeates the vacuum. So if this picture is true, then why isn't the universe completely filled with electrons, photons, and other particles everywhere? Well, one immediate surprise is that the universe can be filled with fields but be absent of particles. Something has to trigger all that potential energy in the fields to become particles, just like a push can make a ball roll down a hill to accumulate kinetic energy.

When we look at the cosmic microwave background (CMB), it is filled with particles. However the lowest energy state in the universe is the vacuum state, and it is also the most symmetric situation that the universe could find itself in. If we go back in time before the CMB existed, the universe was a vacuum, and somehow particles emerged from this vacuum. We discussed that the time energy uncertainty was at work to create particles out of the vacuum during the early stages of the universe. But this vacuum fluctuation would have created both particles and antiparticles, and they would have quickly annihilated each other. The vacuum fluctuations are not enough to keep our particles around for a long time.

For the current observed particles—present in both galaxies and our very bones—to come into being, vacuum fluctuations needed to create more matter than antimatter. For this to happen, baryogenesis was needed: a hypothetical process that created asymmetry in the universe so that we wound up with more matter than antimatter.

As discussed before, we saw that cosmic inflation preceded the radiation-dominated epoch when the CMB existed. Inflation is a flash of time in which no particles of the standard model exist. If inflation is valid, then it must play a crucial role in generating the observed particles that constitute the structure of our universe today. As I write, there is no consensus on what the correct theory of baryogenesis actually is and when it specifically occurred. There are a few proposals on the table, and they fall into three basic categories:

1. Baryogenesis takes place during inflation.
2. Baryogenesis takes place right after inflation.
3. Baryogenesis happens during the epoch when the electroweak interaction is dominant.

It turns out that regardless of when baryogenesis happens, there are some universal features that, if satisfied, will produce the observed matter in the universe. These conditions were named after the great Russian physicist Andrei Sakharov. Sakharov was one of the main players in the Russian nuclear arms program and is often called the father of the Soviet hydrogen bomb. But he later renounced the proliferation of nuclear arms and became an international voice of pacifism and defender of human rights in the Soviet Union, earning him the Nobel Peace Prize. Instead of research in nuclear arms, Sakharov turned his attention to the origin of matter in the universe and was the first to propose the necessary conditions for the early universe to generate matter from the vacuum. Essential to his arguments for baryogenesis are the violation of three types of symmetries. Before discussing Sakharov's conditions, it is useful to turn to an important musical analogy, which will help clarify the physics of baryogenesis—resonance.

Recall that to get resonance an oscillating external force is applied to a material with a natural frequency. The amplitude of the vibrating material will rapidly grow if the external force oscillates at the natural frequency of the material. More complicated objects like strings and musical instruments accommodate a wide range of natural frequencies. It is therefore possible to generate a wide range of resonant frequencies with an external force. Quantum fields are like an extended material and, like a string, can vibrate with many resonant frequencies. Because all the quantum fields can interact with each other, one quantum field can behave like the external force, and the other quantum field can resonate in response to this interaction. A quantum field that sits in a vacuum can be driven to oscillate at a frequency that equals its rest mass-energy. If we use Einstein's relation

$$E = hf = mc^2$$

we find that driving frequency, f, of one quantum field on another with mass, m, can resonate a quanta of vibration if the driving frequency is $f = \frac{mc^2}{h}$. Therefore particles are resonant vibrations of quantum fields and can be realized similar to how musical notes are created by plucking a guitar string. The plucking is the external force from the interacting quantum field, and the particle created is analogous to the note generated. But to apply this analogy to the real universe, the vacuum instrument needs to be rigged a little so that some vibrations are not allowed, such as antiparticles, and this is connected to the breaking of certain symmetries in the vacuum.

Now let us learn from Sakharov what types of symmetry of the vacuum we need to break. When we look closely to the symmetries of the standard model of fundamental interactions—which means all inter-actions between bosons and fermions—we see something striking: all these interactions are symmetric under a combination of reversing the order of time, changing the spatial orientation of the interaction (like looking in the mirror), and reversing the charges of the particles. Let's consider the case of the electron and the positron in the Feynman diagram. First, change the sign of a particle's charge, then reverse the time order, and finally reflect the diagram from left to right—a mirror reflection. Amazingly, the diagram will describe the same physics, except that the electron and the positron travel backward in time to produce photons. This has been tested in particle accelerators with hairbreadth precision and found to be true. What Sakharov found was that a combination of these symmetries had to be broken in the early universe. Even more exciting, a new physics is needed to break these symmetries.

The first symmetry that had to be broken is called the baryon current—similar to an electric current in a wire, except that baryon current can exist in space. In the standard model's vacuum, the rate of the baryonic current is exactly zero. In other words, there is an equal

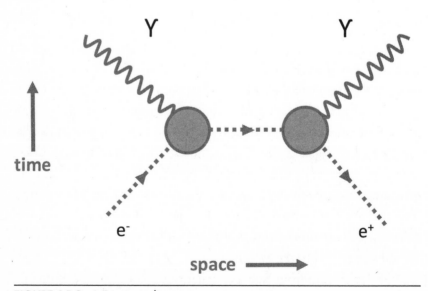

FIGURE 15.2. A Feynman diagram.

likelihood for the baryon current to flow into and out of the vacuum sea—a symmetry between these two processes. However, some other field that resonates with the baryon current field could violate the symmetry in the baryon current: let's call this B (for the baryon symmetry). But this is not enough to get baryogenesis. Whatever violates the baryon current will also violate the current for antibaryons because the vacuum also has a symmetry between baryons and antibaryons (the symmetry between matter and antimatter that Dirac discovered). If antibaryons are produced, they would annihilate the baryons that are created, leaving us with nothing. So the symmetry that relates a particle to an antiparticle—let's call it C—in an interaction had to be violated as well. And finally if, using that new physics, we simultaneously violate B and C, these created baryons over antibaryons need to be protected from reaching thermal equilibrium with the background radiation, otherwise our newly produced baryons would melt into the heat bath. A handful of baryogenesis models are on the market, but all of them are proposed to take place after inflation happened. Perhaps a universe of sound could point to a baryogenesis mechanism that arose from a field that drove a resonance, which could satisfy all of Sakharov's conditions in one fell swoop.

In 2005, I was reluctantly walking to work at the Stanford Linear Accelerator Center (SLAC), where I was continuing my postdoctoral work, frustrated that no new ideas had arisen for months, when I spontaneously remembered a dream I had had a year earlier, while at Imperial. I had related my strange dream to Chris Isham. In the dream, an old man in a white robe in outer space was writing some equations at a lightening fast rate. Frustrated, I had pled with the man that I was too dumb to comprehend the equations. Then the blackboard disappeared, and the old man slowly swirled his hands in a spiral in one direction. I had thought nothing much of the dream at the time, but Chris had kept asking me details about it, such as which direction the man was swirling his hands. The theory group at Stanford was fixated on cosmic inflation and its possible inherent symmetries. I now realized that the orientation of the man's swirling hands provided an insight into how to break the symmetry of cosmic inflation and generate the baryon asymmetry. As I walked down Palm Drive, my grimace turned into a big smile. I saw an opportunity for getting baryogenesis from cosmic inflation.

I felt so happy that I went to go have a beer in a café across from the computer science department. Then Michael Peskin, my postdoctoral advisor and the head of SLAC's theoretical physics group, came walking by with one of the star postdocs. I waved to Michael and told him that I had made a connection with the broken symmetries of the vacuum and baryogenesis. The other postdoc smirked and said, "Another one of your crazy ideas." But Michael always would give a theorist an opportunity to make his bed and sleep in it. We made an appointment to talk.

In the theoretical physics community, Michael Peskin is known as "the Oracle." An unassuming man with a moustache and specs, he possesses an encyclopedic knowledge of physics, especially quantum field theory and supersymmetry. Most of the postdocs, including myself, were scared to talk to him, not because he was mean but because you would always walk away from a discussion with him eating a physics humble pie. I first met Michael during my last year of graduate school in a string theory summer workshop, where he gave some lectures on the standard model of particle physics. There is an insider joke among some theorists that Michael is the Lt. Colombo of physics. For readers who aren't that old, *Colombo* was a popular eighties TV show, featuring detective Colombo. The disheveled detective was polite to a murder suspect and came across as naïve, but these mannerisms were a distraction. Colombo would ask what appeared to be dumb questions that would both irritate the suspect and trap him into making inconsistent statements and an ultimate confession.

The Stanford Linear Accelerator is famously known as one of the hardest places to give a theory seminar. I often witnessed in the middle of the seminar this unassuming man politely raise his hand and say, in a sincere high-pitched voice, "Excuse me, but I'm a little confused."

FIGURE 15.3. The author's postdoc advisor at Stanford SLAC, elementary particle theorist Michael Peskin. *Michael Peskin*.

The speaker would fall for the bait, initially feeling sorry for the poor confused Peskin. Then—*bam*—the temperament of the speaker would go from pity to terror once he or she realized that the poor "confused" soul had destroyed the speaker's entire talk. So imagine being Peskin's postdoc and having an office next to his for three years!

When I spoke to Michael about the idea, he actually liked it and told me to do a long calculation: shut up and calculate. Then I realized that the calculation would actually take months to perform. I teamed up with a fellow Stanford postdoc Shahin Sheikh-Jabbari, a brilliant string theorist from Iran. I was looking for a permanent faculty job that year, so I was desperate to finish this project. Every time Shahin and I would go back to Michael with some progress, convinced that the project was finally reaching a conclusion, Michael would say, "Oh, I'm so sorry, Stephon, but I'm confused." This went on for eleven months. Time was running out.

After months of work we demonstrated that the inflaton field, to our pleasant surprise, broke the CP symmetries of the vacuum. And the inflaton did it with resonance at the heart of it. Since space-time can bend and stretch, Einstein showed that a disturbance of space-time can ripple a wave of gravity that moves at the speed of light—a gravitational wave. Cosmic inflation generically produces a gravitational wave because the inflaton field, like a stone on a pond, disturbs the space-time fabric.

In most inflationary models, two types of gravitational waves are produced, one that spins in a left-handed manner and one that spins in a right-handed manner—the same way a football spins differently if you throw it in the same direction with your left hand or your right hand. But what we discovered for the first time was that during inflation, the inflaton field resonates a left gravitational wave with much greater amplitude than a right-handed one. And it turns out that left-handed gravitational waves interact uniquely with matter, and right-handed ones interact with antimatter. As a result, the greater amplitude in the left-handed gravitational wave resulted in the resonance of matter over antimatter. This situation created a simultaneous condition of CP violation and baryon number production from the production of gravitational waves—satisfying two of the three Sakharov conditions from the same agent, the inflaton. And finally the last Sakharov condition—out of equilibrium—occurred naturally because during inflation space was expanding much faster than the baryons were being created.

Michael's constant and often frustrating confusion forced Shahin and me to dig deep. Michael knew exactly the weak spots to target, which allowed us all to unveil the new discovery of baryogenesis during cosmic inflation. Talk about practicing one's chops and stumbling to make that brilliant solo. Our new baryogenesis mechanism was interlinked with how the quantum fluctuations of the inflaton field play a role in initiating the structure of the universe and the origin of matter over antimatter. How inflation actually does this quantum dance requires a special kind of song and opens a Pandora's box.

16

THE BEAUTY OF NOISE

Complex structures such as stars, galaxies, and planets arise from sound waves in the primordial plasma. In producing these sounds, the universe functioned like an instrument. But does our analogy with music stop simply at the realization that the universe made these sounds? After all, it seems intuitively obvious that music is more than just sounds: to human ears, periodic waves are the most musical, the most pleasant. However, some composers like John Cage beg to differ:

> I think it is true that sounds are, of their nature, harmonious . . . and I would extend that to noise. There is no noise, only sound. I haven't heard any sounds that I consider something I don't want to hear again, with the exception of sounds that frighten us or make us aware of pain. I don't like meaningful sound. If sound is meaningless, I'm all for it.[1]

Noise is usually perceived as an unwanted signal, a sound to be eliminated. Good headphones are those that reduce noise. Unwanted noise happens even in science: Penzias and Wilson were desperate to get rid of the noise confusing their coveted radio signals. Ironically, the "meaningless" noise they were struggling to get rid of was the cosmic microwave background radiation! Perhaps, like John Cage, we should listen for the music in noise.

We can understand how noise can arise using the Fourier idea. Simply add waves of all frequencies, each with the same amplitude, and we end up with a featureless white noise signal. Our ears will perceive this sound as a hiss because there is no one frequency that dominates: each contributes to the sound equivalently. So white noise is actually the most democratic sound.

As it turns out, cosmologists are on a hunt for the theory of the early universe that generates *noise as the foundation* for structure. The trouble is that we don't know how the first vibrations that led to the primordial plasma began from a state of nothingness. We do not have complete experimental confirmation of the origin of the "first sound." There are a few compelling theories on the market, but all of them need a lot of taming to even stand up to current experimental results let alone to make new predictions. Consider the so-called fine-tuning problem: the essence of it is that we don't know why the constants in nature take on the particular values in our universe that they do. Why is the value of the speed of light what it is? Why are the coupling constants, dictating the strengths of particle interactions, just so? Our universe resembles a finely tuned instrument. Observations are ahead of the game. By charting the rotation curves of galaxies, we have indirect evidence of the existence of dark matter, for instance. It is elusive, but without it, galaxies couldn't have formed. We have subsequently come up with models for dark matter, but which is the correct one? Greater still, what is the correct theory of the early universe?

As it stands, cosmic inflation is our best theory for the early universe because in one shot it is able to resolve two important problems that the standard big bang theory suffered from. The first, which we've discussed, is the horizon problem, which questions how the photons in the cosmic microwave background (CMB) could possibly have the same temperature when they did not have time to interact with each other in the past. As we've discussed, inflation solves the problem by providing an era during which parts of the universe, now too far apart to interact, could interact. The second problem is in finding the correct physics to generate the vibrational energy necessary to source the primordial plasma sound.

We can begin to find the answer by digging into how inflation works in a little more detail. In general relativity, the gravitational field, or space-time, can be thought of as an elastic medium. When matter interacts with the gravitational field, it can alter its elasticity. Alan Guth, the father of inflation, realized that if, during the early universe, a strange substance with negative pressure dominated space-time—essentially, energy that creates repulsive gravity—then the universe would have undergone an expansion so fast that it would be superluminal-faster than light itself. This theory does not, however, violate Einstein's relativity; though relativity restricts objects within space-time to traveling no faster than the speed of light, relativity does not limit the rate of expansion of space-time itself.

Inflation is fascinating: although it is a theory of cosmic expansion, it is rooted in quantum field theory and symmetry arguments as well. Alan realized that a very symmetric early universe is not stable. It's like a pencil balanced, tip down, on a table. A pencil like that has a lot of rotational symmetry, but the tiniest disturbance will make it fall in any given direction, which breaks the original rotational symmetry of the upright pencil. This idea, called spontaneous symmetry breaking (SBB), is ubiquitous across all realms in physics. Recall the Ising model of magnetism that I worked with in Leon Cooper's lab. When the temperature in a metal is nonzero, individual spins are randomly pointing in different directions, and the magnetization is zero. The average orientation of the magnetic field is spherically symmetric. But the nearest neighbors in a ferromagnet prefer the magnetic fields to be aligned in the same direction. So, when the temperature falls to zero, the energy is minimized for spins pointing in one given direction, and the symmetry is broken. The lowering of symmetry as we lower temperature is an example of symmetry breaking.

In quantum field theory, a similar phenomenon occurs. Only in this case it's not the temperature that's controlling the breaking of the symmetry but a field that cosmologists call the inflaton, the thing Alan Guth told me about back in college when I asked him if inflation could do work. In regions in space where this inflaton field is nonzero, symmetry is broken, and the negative pressure carried by the inflaton field causes this region of space-time to "inflate" exponentially. That's Alan's inflation.

As the universe expands exponentially, something magical happens. We learned that the quantum oscillator could never be at rest due to the uncertainty principle. The quantum of the inflaton field behaves like a very large collection of oscillators, much like a vibrating string. However, there is a major difference to keep in mind, too: when we bow a violin string, it is possible to generate a series of harmonics, but not all of them will have the same amplitudes. Inflation turns the universe into a special type of instrument in which the majority of the inflaton field's quantum modes are created (excited) from the vacuum with the same loudness. This happens because the inflating background democratically acts like a source for all modes. But this is not enough to generate the initial structure in the universe—it's too symmetric. We need to get rid of some of the symmetry.

Imagine an orchestra of one hundred violins, each playing different notes. If each violin player played his or her respective notes with exactly the same volume, the result would sound like the noise you hear when a radio is in between two stations. The extreme version of this type of sound is simply white noise. This is what the quantum waves would sound like during inflation if your ears were around and able to hear them. But inflation does something even more remarkable. Each wave comes with a phase, which is a number that can create a shift in position or a time delay in a waveform. In principle, there is no reason for these phases to be the same. This is like randomly dropping a bunch of stones in a pond at different points on the pond. The phases will likewise be random, so that when the waves interfere, they will cancel out. But inflation *synchronizes* these phases. It is like having all those violins, which began playing a cacophony of different notes, converge on exactly the same note because some field—the "conductron"—swept over them. In this sense, the inflaton field is our cosmic conductor.

In the 1980s researchers showed that inflation predicted that the initial waves had the correct characteristics to initiate the sound waves in the primordial plasma—otherwise known as a nearly scale invariant power spectrum of the inflaton's quantum fluctuations. The power spectrum is simply a curve that characterizes the loudness of a continuous range of frequencies, no different from how the Fourier idea worked to construct a complex function as a collection of pure waves of different frequencies. At that time there were no observations to either confirm or refute this prediction. Then in 1992, around the same year that Alan Guth visited my sophomore class at Haverford, the COBE satellite measured these fluctuations in the primordial plasma and found the spectrum that inflation predicted. More than twenty years of even more precise measurements of the CMB, such as the Wilkinson Microwave Anisotropy Probe satellite, co-led by my Princeton colleague David Spergel, and the Planck space observatory, and so far all of these observations are consistent with the predictions of inflation.

According to inflation, the big bang is replaced by a small piece of space that is dominated by the inflaton field. This miniverse became large in a few mind-blowing seconds. In a universe that is fourteen billion years old, or 14×10^9, this miniverse is unfathomably small. The quantum vibrations of the inflaton field get stretched along with the rapid expansion of space and provide the seeds for the sound waves that arise in the CMB. Eventually, the inflaton loses its energy and begins to settle. Like an oscillating spring coming to a stop, with small residual oscillations around its resting point, the inflaton field continues to oscillate with its remaining energy. These oscillations will excite the matter fields the inflaton interacts with at this stage and produce particles. In fact, Robert Brandenberger and Jennie Traschen were among the first cosmologists to show that this phase, which they dubbed "preheating," generated explosive production of the standard model.

So inflation uses both general relativistic and quantum mechanical effects to produce the white noise pervading the early universe, which seeds cosmic structure. When inflation ends, this noise, like blowing a cosmic mouthpiece, gets transformed into sound waves in the primordial plasma, exciting particle production. It has been over thirty years since the discovery of inflation, and it is surprising how difficult it has been to come up with alternative theories to generate noise. And alternative theories are needed because, of course, for all its successes, inflation is not perfect.

Inflation's first major problem is its supposition for how loud the primordial sound would have been. When we look at the measurement of cosmic sound's loudness in the CMB power spectrum, we see that it deviates from the average energy of the plasma by a fraction of 1/10000. Any tiny departure from the observed loudness would render the universe an uninhabitable place, since structures would either form too fast, not leaving life enough time to evolve, or not form at all. In its simplest form, unfortunately, the model of inflation (described by a quantum field theory of the inflaton field with zero spin) cannot give the correct value of the white noise's loudness,[2] nor can the inflaton field by itself. The theorist has to introduce a free parameter that controls how the inflaton field couples to itself.[3] In the simplest model of inflation, the self-coupling has to be adjusted by hand to be as small as one part in a trillion. Any deviation from this ridiculously small number will predict an uninhabitable universe. There are literally hundreds of models of inflation, and despite their diversity, they all suffer from similar fine-tunings. What's even more troubling is that the theory gives no insight as to why this tuning should be so small. Quantum field theory, which describes the three fundamental interactions of the standard model (electromagnetic force, strong nuclear force, weak nuclear force), also has fine-tuning problems. One interesting fine-tuning problem concerns the parameters that control the relative strengths of the electromagnetic and strong nuclear forces. If these parameters were any different from what is observed, then stars would not be able to make carbon, which is central for life.

A second major problem for inflation concerns how it began. Once again, it is a fine-tuning problem. What set of conditions enabled the inflaton field to have the correct properties—including an early universe dominated by negative pressure and fields with the correct potentials—to ignite the correct amount of inflation? Soon after inflation was discovered, Alexander Vilenkin showed that if the universe was completely described by quantum mechanics, then inflation could begin from pure vacuum energy, or what he called a state of nothingness. The quantum state of the universe would begin in a vacuum with no space-time and would spontaneously undergo a phenomenon known as quantum tunneling, which would leave it in an inflating space-time. According to Vilenkin's work and that of other physicists, this tunneling phenomenon could happen more than once, giving Vilenkin's initial quantum state of the universe the opportunity to explore many different space-times, each with its own values with various "fine-tuned" constants. In this scenario, we happen to live in one of the larger universes with the right coupling constants—simply because we are alive to make the

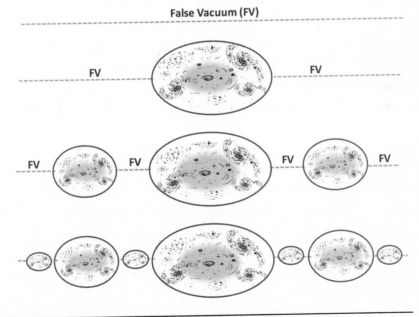

FIGURE 16.1. This figure depicts a proliferation of inflating universes from false vacuum energy. Where there is false vacuum energy, bubble universes will eventually form. This process continues eternally into the future.

measurement. This line of reasoning is called the anthropic principle. Invoking the anthropic principle is perceived by many physicists as unscientific because it is not falsifiable. We cannot observe any other universe—or so most believe. There have been some cosmologists, such as Matt Kleban from New York University, who have postulated that our early inflating universe could have collided with another bubble universe, which would have left an imprint on the cosmic microwave background radiation that we could possibly detect.

String theory, which proposes to unify all four forces, has the ingredients to assign the values of the coupling constants by having extraspatial dimensions. My colleague Lee Smolin in his book *The Life of the Cosmos*, after discussions with Harvard string theorist Andy Strominger, foresaw the challenge for string theory to uniquely determine the couplings of nature due to the many ways the ten dimensions described by string theory can curl up to give rise to four-dimensional space-time. Lee argued that these uncountable possibilities that string theory gives to our four-dimensional world would provide a huge "landscape" of coupling constants, and it would be nearly impossible for the theorist to predict from string theory why we live with the observed constants of nature.

In 2003 when I was a postdoc at the Stanford Linear Accelerator Center, Leonard Susskind put out a paper that would change the whole program of string theory. It was entitled "The Anthropic Landscape of String Theory." The abstract states "whether we like it or not, this is the kind of behavior that gives credence to the Anthropic Principle." The "this" in Lenny's statement refers to his observation that instead of giving one unique solution with our observed coupling constants, string theory admits many four-dimensional worlds with their own coupling constants—consistent with Smolin's expectation. Lenny reasoned that string theory could generate many inflationary bubbles populating the landscape and realizing different coupling constants. I immediately knew the consequences for young postdocs like myself after the master speakth! Our Diracian dreams overnight became more like a nightmare. What is left for us to calculate as the hunt to find that unique solution of our world from the theory of everything seemed futile?

Because Lenny had always been kind to me and was supportive throughout my career, I felt comfortable asking him, "Lenny, now that we are faced with the landscape, what is left to do?" Lenny replied, "We're lucky to be in one of those universes, but we still need to find an example in string theory that realizes inflation." Despite my respect and adoration for one of my biggest heroes in physics, I just could not accept the multiverse idea. One of my biggest objections was the inability to make predictions in string theory about what the fine-tuned values should be. So in the spirit of Ornette Coleman, I changed chords and started to work on other approaches to early universe cosmology.

I have spent the last fifteen years working on both inflationary models and alternative theories to inflation. All of these models have some type of fine-tuning issue. So any insight into the correct physics of the early universe will require us to deal with fine-tunings head on. If inflation or any alternative had any hope of success, then understanding how the constants of nature arise seemed to be necessary.

I serendipitously participated in an intense three-month M-theory boot camp in Paris at the Institut Henri Poincaré. Despite the unpopularity of American politics in Europe in 2000, I made many European physicist friends, especially Chris Hull, who was one of the main lecturers at the workshop and copioneer of M-theory. Along with Paul Townsend, Chris realized string theory enjoyed even more symmetries, or automorphisms, than the ordinary field theories that describe point particles. In 1995 Witten gave a historical talk at the University of Southern California and, based on Chris's calculations, conjectured that all five string theories are actually different manifestations of an underlying eleven-dimensional theory. Another key idea of M-theory is that it is not only a theory of strings but other higher dimensional vibrating objects called D-branes.

Joe Polchinski discovered that strings could collectively end on these surfaces called a brane of varying dimensions. The simplest case to visualize is a two-dimensional surface, or a 2-brane. This is possible because the endpoints of strings are allowed to fluctuate on the brane's surface,

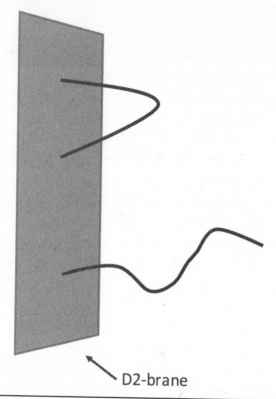

D2-brane

FIGURE 16.2. A D-brane of particular interest is the D3-brane, a three-dimensional membrane. Inside this D3-brane, the fields of the standard model, which are related to the oscillations of the strings ending on the brane, are confined. So a well-dressed D3-brane can be a good candidate for our universe. This was not a new idea as it had been previously formulated by Lisa Randall and Raman Sundrum.[4]

as opposed to being fixed, like a guitar string. To everyone's delight, Joe realized that these surfaces are physical objects because they solve two longstanding puzzles in string theory. First, strings are allowed to be stretched (open) or closed into a loop and only the closed string enjoys the target space duality. By including the D-branes, Joe showed that the open strings also had target space duality. Second, the quantum theory of strings had another charge, the Ramond-Ramond charge, with no identifiable object to which it couples. Joe showed that D-branes were the objects that carried the Ramond-Ramond charge the same way a point particle (0-brane) carries electric charge.

At that time I was trying to find an intuitive and mathematical way to realize inflation in string theory, but of course, I wasn't the only one. Despite my learning a lot of string theory, the other postdocs were intimidatingly more prepared than I was. Slowly, I began to ooze out of lectures and physics discussions and found myself blending more naturally into the local Paris jazz scene. Eventually, I totally disappeared from the prestigious M-theory workshop. I figured that if I was not going to make it in physics the traditional way, then I would have to find my own way. When I was not playing, I would scribble diagrams and inchoate equations on damp Nutella-smudged napkins. Then I'd swap over to writing down mnemonics for playing through the chord changes of a standard I was working on. My music breaks were accompanied by cheap, yet enjoyable, local house red and Nutella-banana crepes.

Then it happened, as I was soloing with the equation of D-branes on a piece of napkin, with jazz in the background, I was interrupted by people clapping. In real time, the motion of the clapping hands merged with the D-branes, and a happy thought came to me: What if colliding D-branes could ignite the big bang? Sometimes cosmologists conflate the big bang with cosmic inflation.

It was no coincidence that I was struggling to understand the new paper by the prodigious Indian string theorist Ashoke Sen on colliding branes. When particles and antiparticles collide, they annihilate each other and produce radiation. But Sen showed that when a brane and an antibrane collide, they annihilate each other but also produce lower dimensional branes. In particular when a D5-brane and an anti-D5-brane annihilate, they produce a D3-brane. I was exploring the physics of D-branes because they had the potential to address the fine-tunings in inflation. D-branes are also powerful because the motion open strings that end on the D-branes generate quantum fields within the D-brane. So our universe, and the fields of the standard model, could exist within a D-brane with three spatial dimensions, a D3-brane. Even more interesting, the coupling constants are controlled by the bending and stretching of the D-brane. After a few more glasses of wine, I went to my closest friend in string theory at the time, Sanjaye Ramgoolam, who was a trailblazer in the field at the time. He might laugh, but he wouldn't judge me. We were pals.

We met at a brassiere in Odeon. "Sanjaye . . . I think I have a way of getting inflation from string theory," I said. I told him the half-baked idea by drawing some pictures. Sanjaye has a way of looking at you with his piercing eyes when he is skeptical but serious. "You know, you always have these ideas, well it's time to put out or shut up . . . Show me some equations, then we can talk." I took this as good news. Sanjaye was very good at immediately demolishing my ideas with tough love, and he tried to in this case, but I kept coming back. Clearly, my years of scuffling with my string theory mentor had made me stronger. I hurried back to my café-office with a bunch of papers and did my calculations. I was on an unspoken high, staying up almost entire nights. In an ecstatic frenzy, I improvised my calculations, with a certainty that the equations would work out. After a few months, I was done. When I presented the calculation to Sanjaye, he simply said, "You nailed it!" In a nutshell, I was able to come up with the first model of inflation based on the annihilation of D-branes.

The key insight, as discovered by Sen, is that an observer living on the D5-brane sees the created D3-brane as a string like vortex, the same way that a three-dimensional observer sees a one-dimensional object as a string. In both cases, the difference in dimensionality of the D5, D3, and D1 is two. So a vortex can be generalized to any dimension so long as the difference in dimension is two. Cosmologists in the pre–string theory days tried to construct inflationary models where inflation took place in the center of vortices, but they didn't work because of fine-tuning. I was able to address the fine-tuning problem with a few "healthy" assumptions and show that inflation could take place in the D3-brane. Recall that one of the problems with fine-tunings was that a coupling constant could not be a priori determined by the theory. In my model, the coupling is determined by a quantity intrinsic to string theory, its tension. I was able to find a mathematical solution of an inflating 3-brane universe that had milder fine-tunings, and with the advantage that the coupling constant can be determined by the theory.

When I returned to London, I shared a draft of my paper with one of the top string theorists on the planet, Arkady Tseytlin, who other physicists refer to as "the human computer." If there was anything not rigorous enough about the work, Arkady would catch it, and I was certain that a solid theory was too good to be true. Arkady conservatively said, "Post the paper." Hallelujah! Two weeks after putting out my paper, a group of theorists at Cambridge, McGill, and Princeton posted a similar idea of inflation based on brane-antibrane interaction. Finally, after all those years of dreaming of one day composing my own theoretical invention, I had done it. The paper entitled "Inflation from D-Anti D brane Inflation" was published in 2001 and enjoys over two hundred citations.

My paper turned out to make a big contribution to a subfield in string theory and cosmology, but it became clear to me and even more seasoned string theorists that the fine-tunings were replaced by other fine-tunings—the need for a multiverse to anthropically select the coupling constants in every universe that was created during many realizations of inflation. It started to feel like an eerie case of Ptolemy's epicycles. I still remain a fan of string theory and the more sophisticated stringy inflation models that my work later inspired, but a new direction in inflation research came from another lucky chance meeting with another jazz physicist.

I met David Spergel originally when I was a postdoc at the Stanford University Linear Accelerator Center. He is one of the lead scientists for the Wilkinson Microwave Anisotropy Probe (WMAP), and one of his subsequent papers resulting from the WMAP satellite stands as the most cited paper in all of physics. David is the head of astrophysics at Princeton and is, simply put, one of the giants of our field. Except, you wouldn't know it. When I first met him, he seemed like one of the guys from around the corner. At the time, David was sporting a hip goatee and was decked out in a colorful Hawaiian shirt, shorts, and sandals. The postdocs were in awe of him, and most of us were scared to talk to him. At the time, I was feeling like an outcast in our theory group. One day, David was hanging out with some professors and postdocs when I nervously approached him. I didn't know what to say, and something speculative and laughable slipped out of my mouth—a crazy idea I was toying around with. To the other professors' surprise, David took my question seriously and suggested that I pursue the project. He had this power to "see" answers before doing calculations, similar to how Einstein came up with *gedankenexperiments*.

Years later, I reconnected with David, and it happened again. I expressed to David my dissatisfaction with inflationary models getting too complicated. We started informally talking about the possibility that inflation might occur from known physics. What's the most established physics that could do the job? We puzzled over this, and then, eureka, the answer was in front of our face. Light! The electromagnetic field carries energy. What if the preinflationary universe was filled with light radiation; perhaps this energy could make space inflate. David invited me to be a visiting professor at Princeton for a year, while I was on my sabbatical from Haverford, and we, along with my postdoc Antonino Marciano, worked out a very simple, Occam's razor model of inflation driven by the interaction with light and electrons. In this case, no exotic physics beyond the standard model was needed—but the universe had to start out in an unusually flat situation. This model represents a new generation of inflationary models that is based on physics that we already know to be true. Much work needs to be done, but the model is a proof of principle that inflation may not rest on exotic theories of quantum gravity.

17

THE MUSICAL UNIVERSE

The fine-tuning problem pervades all realms of physics. When I think about the epitome of the jazz physicist, the image of João Magueijo comes to mind. His improvisational training came early in life as an avant-garde classical pianist-composer and a black belt in shotokan karate. I first met João when Robert invited him to Brown to give a seminar. In preparation, our group had to discuss what first appeared to be an iconoclastic paper written by João and inflation pioneer Andy Albrecht. The paper was entitled "A Time Varying Speed of Light as a Solution to Cosmological Puzzles." Oh, come on, didn't Einstein make it clear that nothing could travel faster than the speed of light? Who was this guy who dared to challenge the great Einstein? João and Andy showed that if the speed of light went to infinity in the early universe, the CMB photons could have time to interact with each other to solve the horizon problem. There would then be a mechanism to allow the speed of light to settle down to the constant value that we observe in the late universe similar to how temperature can control the onset of magnetism in a magnetic material. What this meant was that there was an alternative to the solution inflation provided for the horizon problem. Plus, it had the potential to help out the fine-tuning problem since it embraced the idea that a theory could have couplings between fields that change over time as opposed to being randomly distributed in multiverses.

Naturally, the question of whether other constants of nature could vary in time arose. Why just the speed of light? In a group meeting, Brandenberger, with his usual open mind toward alternative theories, suggested that fundamental constants need not always be universal in a quantum theory of gravity. He was right. Einstein's theory says that light travels at a constant maximum velocity in *empty* space. It is the mathematical symmetry of special relativity, called Lorentz symmetry, that preserves the speed of light within frames of reference that move at constant velocities with respect to one another. The idea of special relativity is that there are many frames of reference, each with observers living within them. The speed of light in each of these frames of reference has to have the same velocity, regardless of how fast those frames are moving with respect to each other. In other words, if an observer looks at another frame of reference (like a moving train), he or she will still see that the speed of light is the same even though that frame of reference is moving relative to it. It is because of this symmetry that our standard theory of light, or electromagnetism, cannot accommodate a varying speed of light in empty space. But when a wave of light moves in a different medium, such as glass, Lorentz symmetry is no longer preserved, and light can change speed relative to empty space. This was the essence of João's contention. It could be that there is a quantum effect on space-time that fundamentally violates Einstein's cherished Lorentz symmetry, resulting in a variation of the speed of light in the early universe. As it turns out, the shape of extra dimensions in string theory can indeed cause certain constants, including the speed light, to vary throughout the space-time fabric.[1]

As we walked into the seminar room for the lecture, João was at the front with a mischievous grin on his face, wearing a plain black T-shirt and black denim jeans. There appeared to be some groupies in attendance to see the handsome jet-black-haired Latin bad boy of physics debate with the older men in tweed. Most of the professors settled themselves at the front of the room, prepared to do battle with him

FIGURE 17.1. Theoretical physicist Joäo C. Magueijo.

in his attempt to stand up to the treasured idea of a constant speed of light. But to everyone's surprise, and likely to his amusement, the trickster didn't talk about his challenge to Einstein's theory.

Joäo's talk, in the field of quantum cosmology, was called "Photographing the Wavefunction of the Universe." The wavefunction of the *what*? Quantum mechanics is normally associated with subatomic matter. But we have seen that the classical laws of the macroscopic world emerge from quantum mechanics—quantum fluctuations provided the pressure waves in the universe, the seeds for all structure we see today. Therefore the philosophy in quantum cosmology is to apply quantum mechanics to the entire universe. This means that in the inflationary epoch not only do we quantize the inflaton field but also space-time itself. Therefore quantum cosmology uniquely spans the realm of any quantum theory gravity.

After his talk, João and I spent a few hours discussing some new ideas to realize his varying speed of light theory in the context of string theory. He reminded me of the jazz musicians Sonny, Trane, Miles, and Ornette who mastered the tradition and used those tools to push the tradition to a new level, without fear of what the authorities thought or what the professional repercussions might be. Sometimes those changes were willful, and sometimes they emerged out of improvisation. In that spirit, I embraced the variable speed of light theory, and at that moment in the café, we improvised physics together. It was a relationship that was to last for a long time, because when João returned to Imperial College in London, he teamed up with string theorist Kellogg Stelle and hired me as a postdoc to work with him on extending his variable speed of light theory to the realm of quantum gravity.

The acts of embracing and improvising have made me who I am in physics. They are two values I learned from people who were masters at both: Kaplan, Cooper, Brandenberger, Jaron, and Coleman. Weaving music and physics into one avenue of thought has showed me how to use notions in music as points of access to various fields in modern physics and cosmology. Analogies have helped make the physics more accessible and stimulating.

It is wonderful to think of following the footsteps of our ancestors— the great ancient thinkers who sought to understand physics through sound, and sound through physics. Pythagoras played with hammers and strings to try to understand where the pleasures of music came from, while Kepler used his intuition that the universe was musical to make major advances in the fields of astronomy, physics, and mathematics.

Music and sound have persisted, whether we have focused on them or not. They are part and parcel with the universe. The symmetry of musical compositions mirrors the symmetry that exists in quantum fields, and the breaking of these symmetries in both cases lends beautiful complexity. In physics we get the distinct forces of nature with symmetry breaking, in music we get tension and resolve.

The uncertainty of being able to know both where a particle *is* and where it is *going* beautifully mirrors jazz improvisation. And isn't it mind blowing that the spectrum of vibrations that were amplified by inflation, those that led to the structure in our universe today, is the same as the spectrum of noise? Fundamental to it all is the Fourier addition of waves. The harmonic structure of the cosmic microwave background emerges from quantum noise, just as distinct beats and rhythms emerge from a fundamental waveform, an oscillation, a uniform repetition, a circle. There's a reason why a Stradivarius violin is coveted by practitioners: they don't make them like they used to. Every instrument has its sound, its character, and the universe is no different. That's exactly why physicists look at the oscillations of the CMB for characteristic traces of dark matter or dark energy. And these oscillations continue to persist in the patterns in clusters and superclusters of galaxies today.

Tell a child that the first stars and galaxies in the universe were created by sound in the primeval plasma, just after the birth of our universe, and that those in turn created galaxies with complex patterns and stars that sing with particular resonant frequencies. And tell them that there is much more to the universe than this. All analogies break down. But as Leon Cooper taught me, a powerful analogy can say something new that you otherwise would not have known by the theory on its own. Teach children these analogies, and they will push the boundries later on.

What happens if we push the analogy between music and physics now? Similar to how Mark Turner's explanation of improvisation helped me understand quantum mechanics, what else can music teach us? What if we are going to turn this analogy between music and the cosmos into an isomorphism—a one-to-one correspondence—and speculate that the universe *is* musical and see what it teaches us. Could it generate new physics or point to a preferred option in a controversial debate in cosmology? Let's explore this, these new ideas, together.

A major contention with the notion that the universe could be musical is the use of the word *music* in the realm of the universe. Music is conceived as a human creation, based on our perception of sound, organizing sounds according to harmony, rhythm, and melody. But music is also about using noise and discord to create rhythm and tension or to change an expected harmonic orientation of a musical piece. When I refer to a musical universal, I refer precisely to these elements but generalized to any medium that supports wave phenomenon, namely physics and the universe. If the universe is musical then it is fundamentally wave-like and can be represented as a temporal evolution of sound waveforms. Or as modern composer Spencer Topel says:

> While it's almost impossible to define music—due to all the different kinds of variations to what people think music is—it can be represented as a complex chaotic waveform that also contains beautiful structures. Likewise, when we look up at the sky, we see beauty as well as chaos. Both the cosmos and music are driven by the relationships and structure of waves. They both contain nearly incomprehensible complexity, yet we can discern structure and make meaning of what we see and hear.[2]

Get this: If there is nothing *outside* the universe and if the universe functions like an instrument, with all the musical elements it has, then the universal instrument must play itself. In other words, the cosmic sound is the instrument and the instrument is the cosmic sound. Everything in the universe, including space-time, that supports it must vibrate or oscillate.

It's possible to convert this idea into a physical one simply by oscillating one parameter—the expansion rate of the universe. If the expansion rate oscillates with a frequency of a pure tone then we have what I will call a rhythmic universe, otherwise known as a cyclic cosmology. It turns out that Einstein's theory of relativity admits a cyclic universe as an exact solution. This type of universe gets us around the mind-bending question of "What happened before the big bang?" The answer is that the universe underwent a succession of contractions and expansions—there was no beginning. There is no big bang singularity,

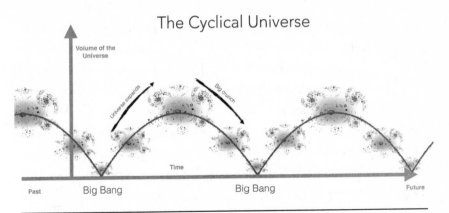

FIGURE 17.2. An illustration of the rhythmic (cyclic) universe.

and time always existed. This is the purest tone that the universe can play. The tone itself is the oscillation of the scale of the universe.

This cyclic solution is actually an ancient idea. One of the earliest versions of a cyclic cosmology came from the ancient Hindu philosophy. In this case, the universe was eternal in time and is created and destroyed in cycles that last for 8.64 billion years.

Coltrane's study of ancient Chinese and Indian philosophy and music brought him closer to current-day cosmology than he probably would have imagined. It is tempting to imagine that Coltrane's improvisation on "Giant Steps," which had a cyclic structure, was embodying the cosmic dance of the expansion and contraction of a cyclic universe. Albert Einstein realized that his theory of general relativity admitted an oscillating universe. In this case, the space-time of the universe underwent infinite past successions of expansions and contractions. Our big bang was one of infinite bangs. Much like inflation, the cyclic universe has its strengths and challenges, and so it is still under active research in cosmology. One of the main challenges of the cyclic universe concerns the existence of a menacing field called a ghost.

In order to get a cyclic universe, the contracting past universe has to emerge into an expanding universe. Cosmologists call this phenomenon a cosmic bounce. Think of dropping a ball. For the ball to change direction, it has to hit the ground, decelerate, come to a stop, and change its downward velocity. This happens naturally due to momentum conservation and the elasticity of the ball. Similarly the "speed" of the decelerating universe will come to a halt and bounce into an expanding state after reaching a vanishing speed. In order for this to happen, we need a field that makes space-time like an "elastic" ball; called a ghost field, this field is an infinite reservoir of negative energy. Physicists do not like ghost fields because they can quantum mechanically transform to an infinite amount of light energy spontaneously. This happens because, according to an exchange Feynman diagram, the photon, the lightest particle in nature, can steal negative energy from the ghost field to create an explosive amount of photons. We don't see such explosive signatures of ghost fields today, so if cyclic universes are for real and do depend on ghost fields, then the ghost fields have figured out a clever way to not decay into photons. Nima Arkani-Hamed of the Institute for Advanced Studies and his colleagues proposed one way this could happen: the ghost field condenses, which bounds the negative energy to a finite value, preventing further decay into photons.[3]

Whether the ghost field turns out to be the real agent behind getting the universe to reemerge like a phoenix from a cosmic death, something as strange or stranger might be needed. In either case, it is still incredible that Einstein's theory encompasses a universe that vibrates like a pure note, a universe that superficially functions simply while still leaving room for so much complexity within. However, it took about seven decades for cosmologists to resort to a cyclic possibility. As early as the 1920s, Albert Einstein considered the oscillating universe as an alternative to the ever-expanding one that his theory predicted. However in 1934, Richard Tolman pointed out an inconsistency with the cyclic model due to the second law of thermodynamics—that entropy will always increase as time progresses. As the universe goes from cycle to cycle, the entropy will increase and the cycles will get longer.[4] Extrapolating back, the cycles shorten, leading to a big bang singularity—no eternal cycles. It's back to the drawing board. But the fine-tuning problems of inflation have given some clever cosmologists the courage to reconsider the cyclic universe.

Despite the reluctance of many cosmologists to touch something as strange as a cyclic universe, the idea of a cyclic universe and a resolution to Tolman's problem was executed by Paul Steinhardt and Neil Turok and revisited several years ago by John Barrow, Dagny Kimberly, and João Magueijo. They were studying whether the coupling strength between two electrons could vary as the universe contracted in to the "big bang" region. It did.

The problem with the multiverse hypothesis is that it is very hard for a theorist to perform a solid mathematical calculation to describe the actual creation of a bubble universe. In the case of the rhythmic universe, the exact solution of a sinusoidal expansion and contraction allows us to get around the mathematical and conceptual issues concerning creating baby universes, which seems to require a full-blown theory of quantum gravity, something we don't have as yet. The musical analogy pushed to its fullest potential paves a new way to solve the fine-tuning problem. And how magnificent that it rests on an oscillation, a circle.

While writing this book, I put myself to the challenge to find an isomorphism between jazz and cosmology. This exercise led me to a new cosmological mechanism for resolving the fine-tuning and getting rid of the multiverse hypothesis. It started with making an analogy with Coltrane's "Giant Steps" and then turning the analogy into an isomorphism. Let's go back to a John Coltrane solo. Coltrane was known to take very long solos—sometimes for hours. The structure of a song like "Giant Steps" has two embedded cycles. The first cycle is a harmonic cycle and the second cycle is a rhythmic cycle. For the most part, there are three tonal centers that move around, like rotating a triangle around the circle of fifths. And this harmonic rotation repeats itself in time. This is the framework through which Coltrane improvised for songs like "Giant Steps." So I imagined that every time the cycle of the song repeats in time, the improviser gets to play new solos, or even permutations based on what they previously soloed. So what if we mapped the notes to the values of coupling constants and the rhythmic cycles to the cyclic expansions and contractions?

A Cyclic Universe Approach to Fine Tuning

Stephon Alexander, Sam Cormack, Marcelo Gleiser[1]

[1]*Department of Physics and Astronomy, Dartmouth College Hanover, NH 03755*
(Dated: July 6, 2015)

We present a closed bouncing universe model where the value of coupling constants is set by the dynamics of a ghost-like dilatonic scalar field. We show that adding a periodic potential for the scalar field leads to a cyclic Friedmann universe where the values of the couplings vary randomy from one cycle to the next. While the shuffling of values for the couplings happens during the bounce, within each cycle their time-dependence remains safely within present observational bounds for physically-motivated values of the model parameters. Our model presents an alternative to solutions of the fine tuning problem based on string landscape scenarios.

FIGURE 17.3. A Cyclic Universe Approach to Fine Tuning.

In order for the coupling constants to be able to change in the context of a cyclic universe, the couplings had to themselves behave like fields, and these fields had to interact with gravity itself. It turned out that string theory naturally had coupling fields that interacted with gravity. And these fields can change during the bounce. As soon as I realized that couplings change during the contraction to expansion phase, I spoke to my colleague Marcelo Gleiser, and he was astonished. "Let's write a paper on this idea," Marcelo said. After playing with the equations of general relativity for a cyclic universe in the presence of varying constants, we found a beautiful picture that gave us insight into how to resolve the fine-tuning problem. The coupling field does not change when the universe is expanding; the energy remains latent as so-called potential energy because it is there, ready to be used—but it isn't. But as the universe undergoes an oscillation from contraction to expansion, the coupling fields gain a lot of kinetic energy. When that happens, the coupling fields are like a ball at the bottom of the hill. Kicked, they can jump over the potential well and change their values. But this kick causes this value to improvise into values that depend a little on the previous cycles. As the universe expands again, the coupling field loses energy and settles back into the well. We were able to show that the motion of the coupling field is random during each bounce. Imagine in the distant past that the universe went through trillions of successions of bounces. During the bounce, the couplings changed randomly. Therefore, we happen to live in the epoch where the couplings have evolved suitable for life. According to this theory, billions of years in the future, the universe will contract, and the couplings will change, and it could be that the future laws will not accommodate life as we know it.

Yet, it is somehow beautiful and satisfying that our lives and all that gave birth to it—our particular quantum fluctuations, our space-time, our stars and planets—might be part of one cycle of a greater whole. Our universe may succinctly be one swing around the circle, waiting to improvise again.

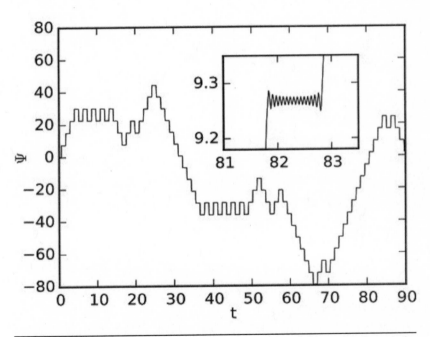

FIGURE 17.4. A numerical solution of how the coupling fields evolve through the bounce during every cycle of the universe. The Y-axis represents the value of the coupling, and the X-axis represents cosmic time.

18

INTERSTELLAR SPACE

It is the knowledge, . . . in eternal truths that distinguish us from mere animals and give us reason in the sciences raising us to knowledge of ourselves . . . The necessary and eternal truths are the first principles of all rational knowledge. They are innate in us. They are the very principles of our nature as of the universe because it is our essence to represent the whole universe.

—Gottfried Leibniz

From the symmetric patterns of quarks that organize to form nuclear matter, to the helical structure in DNA, to patterns of galaxies in super-clusters, the universe is teeming with structure. Even the physical laws that function to create this myriad structure has a structure, which is governed by a continual dance between symmetry principles and their violations. In this book we took a journey through sound, making the case that the unfolding of these structures in the universe has a musical character. The dance among harmony, symmetry, instability, and the gaps of improvisation—all cooperate to sustain cosmic structure. It is as if the unfolding cosmos transpires like a John Coltrane solo. The catalysts of cosmic structure are quantum fields, with their proclivity to harmonize throughout the cycles of expansion and contraction of space-time. The initial vibration of these fields sonified throughout the space-time background, like the vibrating body of an instrument that initiated the first structure in our universe. It is through vibration, resonance, and interaction that the microworld is linked to the macroworld.

Albert Einstein said it right: "The most incomprehensible thing about the universe is that it's comprehensible." How is it that through the laws of physics, stars, planets, and, ultimately, life-forms who could figure out those laws came into being? As we ponder the link between sound, improvisation, and the formation of structure and the causative link with a most interesting structure, life itself, one can't help but wonder: Did the universe create structure for a purpose? We run into murky waters when physicists start talking about purpose. But here I am writing this chapter, and here you are reading these words. Is it not part of being human to seek purpose? After all, we are the product of billions of years of structure formation. Music may likely be the best example of a human endeavor that, on one level, has physical and mathematical roots and, on another, has the ability to evoke such powerful emotions and sense of purpose. It is amusing to speculate that the reason why music has the ability to move us so deeply is that it is an auditory allusion to our basic connection to the universe. If our cosmic origins are seated in sound patterns, is it too far-fetched to think that music viscerally enables us to tap into those origins?

We saw that the spectrum of sound patterns after the big bang relies on finely tuned "constants" of nature, and like trying to balance a pencil on its tip, these constants need to be finely tuned by unknown laws for cosmic structure leading up to life to exist. With the analogy of our universe functioning like an instrument that can tune itself to play the cosmic composition of stars, galaxies, and eventually life, it must have a means to achieve this self-tuning. By taking the musical universe analogy seriously, I proposed a harmonic or cyclic universe as a potential solution to this fine-tuning problem. We saw that if the universe underwent infinite successions of expansions and contractions, like a pure tone, then the constants of nature can self-tune by improvising new values during the bounce between a contraction and expansion. When the universe emerges into an expanding epoch (like our present epoch), the coupling fields get fixed to the final value they attained during the bounce time. If our universe underwent many cycles in its past, then the couplings will improvise new values that will eventually be suitable for carbon-based life. But our question still remains: What is the purpose, beyond the emergence of life, behind the development of structure in our universe? For the rest of this chapter I am going to propose a thought experiment to answer this question. And John Coltrane and his mandala will be the subject of this thought experiment.

John Coltrane was known to practice so much that he would fall asleep with his mouthpiece in his mouth. This practice was guided and fueled by an unquenchable search for meaning in the cosmos. In his later years, Coltrane used his instrument as a tool to search for new connections between music and the universe itself—much like how physicists use experimental instruments to do the same thing. For example, Coltrane went beyond exploring myriad ways to play through the II-V-I chord progression, as was solidly demonstrated in his album *Giant Steps*. Like Eno, he was using sound and music to unravel eternal truths about the universe. He expanded the two-dimensional space of tone and time into a hyperspace that included sonic manipulations such as multiphonics—playing simultaneous overtones—and sheets of sound.

One of Coltrane's biggest idols was Albert Einstein, and he embarked on a multidisciplinary investigation in a fervent search for connections among modern physics, cyclic time in Eastern philosophy, Western harmony, and African polyrhythms. Like Einstein, whose discoveries in physics were greatly influenced not only by physicists but by other disciplines, Coltrane realized that he had to go beyond the Western and classic jazz idiom to make his music cosmic and express the cosmic through music. Coltrane should serve as an inspiration especially for the scientist. I am going to make the case that through independent study, Coltrane acquired a main lesson from Einstein's invariance principle and integrated it into his music. What we will see in Coltrane's mandala is at the heart of the jazz of physics: a jazz musician employing the methodology of a theoretical physicist as a *gedankenexperiment* and strategic tool for improvisation.

Among Coltrane's last three recorded albums were *Stellar Regions, Interstellar Space,* and *Cosmic Sound.* The inspiration for *Interstellar Space* was Coltrane's study of Einstein's theory of general relativity and the expanding universe hypothesis. He correctly realized that the expansion is a form of antigravity. In jazz combos, the gravitational pull comes from the bass and drums. The songs in *Interstellar Space* are a majestic display of Coltrane's solos expanding away and freeing themselves from the gravitational pull of the rhythm section. Coltrane believed that the complexity of the cosmos flows into the actions of humans, and he practiced endless hours to be a conduit of this cosmic force. In his song "Jupiter," one can hear Coltrane literally channeling the orbits of Jupiter's moons in his improvisation.

I remember talking with Coltrane's son a few years back, Ravi Coltrane, at Wayne Shorter's seventy-fifth birthday party. I mentioned to Ravi that I was exploring a connection between his father's music and Einstein's theory of relativity. Ravi gave me a serious look and said, "My father was *deeply* into mathematics and physics." What had driven Coltrane's intuition into and fixation on the cosmos?

I had the good fortune to interview renowned composer and multi-instrumentalist David Amran, who had conversations with Coltrane about his interest in Einstein's theory of special and general relativity. They met in 1956 outside Café Bohemia on Barrow Street in the West Village. David had finished a set with Dizzy Gillespie and joined Coltrane, who was sitting outside eating pie.

He said, "How are you?" I said, "Everything's fine." And then he said to me, "What do you think about Einstein's theory of relativity?" I don't think he was so interested in what I knew about it; I think he wanted to share what he knew about it. I drew a blank, and he went into this incredible discourse about the symmetry of the solar system, talking about black holes in space, and constellations, and the whole structure of the solar system, and how Einstein was able to reduce all of that complexity into something very simple. Then he explained to me that he was trying to do something like that in music, something that came from natural sources, the traditions of the blues and jazz. But that there was a whole different way of looking at what was natural in music.[1]

Even those who understand the mathematical intricacies can easily miss the kernel of Einstein's theory: the elegance of a physical theory that contains and relates more complicated laws from a simple principle. In the case of special relativity the "something simple" was the invariance of the speed of light. By invariance we mean a transformation that leaves a quantity unchanged. For example, I can rotate and transform a point on a bicycle wheel to another point, but the spoke will not change its length. There is a deep connection between

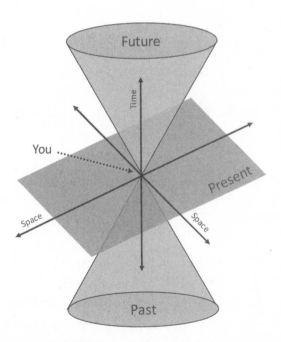

FIGURE 18.1. The space-time geometry of the four-dimensional hypersurface of Minkowski space. Transformations of points on the surface of the cones leave the speed of light the same or invariant.

invariance and symmetry. The wheel has circular symmetry; therefore, a rotational transformation of the wheel preserves how the wheel looks. Similarly, the invariance of the speed of light also reflects an underlying symmetry of space-time. No matter what complicated state of motion an observer is in relative to another observer in space-time, the speed of light is *constrained* to be invariant (constant).

Once this principle is mathematically instantiated, it naturally follows that both electricity and magnetism are unified. All the apparent complexity that appears in the disparate equations gets unified into a simple set of equations that respect the invariance of the speed of light. It's worth showing this. Consider the four Maxwell equations:

$$\nabla \cdot \mathbf{E} = \frac{\rho}{\varepsilon_0}$$

$$\nabla \cdot \mathbf{B} = 0$$

$$\nabla \times \mathbf{E} = -\frac{\partial \mathbf{B}}{\partial t}$$

$$\nabla \times \mathbf{B} = \mu_0 \mathbf{J} + \mu_0 \varepsilon_0 \frac{\partial \mathbf{E}}{\partial t}$$

FIGURE 18.2. The four Maxwell equations of the electric and magnetic fields.

But once we invoke the invariance of the speed of light, all four equations can be written as one master equation: $\dfrac{\partial F_{\mu\nu}}{\partial x^\mu} = J_\nu$

It's worth saying a few words about the master Maxwell equations. The unification of space and time into a four-dimensional space-time continuum allowed Einstein to construct a field that lived in four dimensions. This field is called the gauge potential, which describes the photon and is denoted as Aμ. From this four-dimensional potential field, we can define both the electric and magnetic fields by taking derivatives. This also allows us to define a four-dimensional derivative of the gauge field $d\nu A\mu - d\mu A\nu = F\mu\nu$.

The index and the ν denote the four space-time directions—i.e., $(\mu = (t, x, y, z))$. From this we can define the four-dimensional derivative, $d_\nu = (\frac{d}{dt}, \frac{d}{dx}, \frac{d}{dy}, \frac{d}{dz},)$.

The right-hand side of the equation contains information of both the electric and magnetic fields, but all grouped into one singular object, $F\mu\nu$, that is called the field strength tensor. The right-hand side, $J\nu$, is called the four-dimensional current. It is similar to the three-dimensional electric current in the ordinary Maxwell equations (Equation 1). So the equation simply says that the four-dimensional field strength tensor is sourced by the four-dimensional current. The three-dimensional projection of these four-dimensional objects gives the different three-dimensional Maxwell equations. In three dimensions where the invariance of the speed of light is not readily manifest, the Maxwell equations are pieces (shadows) of a four-dimensional object with manifest invariance of the speed of light. This is like the shadow of an upright bicycle wheel cast on the ground, which can look like a line—the circular symmetry is no longer manifest. Coltrane got this! Based on his words to Amran, I believe that he wanted to do the same for his music. And I will provide some evidence for this.

Einstein's use of symmetry is to constrain how fields in space-time can interact. For example, in the case of electromagnetism, only four-dimensional fields that are constrained to move around a four-dimensional light cone are allowed to interact as seen in Figure 18.1. A good way to visualize this is to imagine being constrained to move on the surface of a sphere with radius denoted by r. If we use the (x, y, z) coordinates to designate a point on a sphere, only values that satisfy $z^2 + y^2 + x^2 = r^2$ are allowed—not any old value of x, y, and z are allowed. We can think of the interactions of fields obeying similar four-dimensional equations that restrict them to live on a four-dimensional light cone. Other interactions that don't live on the light cone are not allowed.

Another role of Einstein's use of space-time symmetry is to dictate relationships between what was once thought to be unrelated phenomena. Before relativity, space, and time—and similarly electricity and magnetism—were naïvely linked. A particle can change its position in time, but relativity relates the very length of space to the duration of time depending on the intrinsic motion of an observer. What looks like an electric field to a static observer is actually a magnetic field to a moving observer.

I am going to argue that Coltrane implemented these ideas of relativity to his music. A revelation of his mandala that I discussed with Yusef Lateef gives us a clue. Just like Einstein's light cone, Coltrane's mandala was a geometric structure that unified relationships between some of the key scales and harmonic devices that he used in his repertoire. Given that musical practice was at the center of Coltrane's musicianship, the mandala could have functioned as a geometric device that revealed a multitude of patterns in the musical universe. Once I realized this, I started to use the mandala as a tool to practice scale relationships between them, guided by the patterns in the mandala.

In special relativity, the fact that the speed of light is fixed causes other quantities to become distorted to maintain the invariance of light in different frames of reference. For example, the length of a train to a moving observer will shrink relative to the same train at rest as seen by an observer.

Likewise, if we were to play the same notes but in two different keys, those identical notes would sound different. Not only would they be perceived as different, but they actually would be occupying different positions in that new scale. Playing A-B-C in the key of C sounds like a sixth, seventh, eighth, ending at a resolved tonic. If I play the same set of notes in, say, the key of B, then they start at the seventh, pass through the tonic, and end at a minor second above the tonic. They stand in entirely different relations to the fixed points (octaves, fifths) of the key they are being played in. We think that those notes, like the length of a train, are a single fixed thing—it is the note A or the note B—but when being played in the context of any given key, they are different, distorted, due to the fixed values of the tonic and intervals within that key. Coltrane's diagram is a more elegant example of this idea where the relationships between the fifths, the tritone, the tetrachords are fixed structures that act as a basis to relate relative scales to each other.

At a first glance the mandala looks daunting, so to find the underlying structure we will reduce it to a skeleton that characterizes the invariance. And like special relativity, once we have the invariant structure, we can generate the complexity in the dynamics from the interactions dictated by the invariance. Our first step is to identify the invariance, or the geometry inviting us to ignore the notes. We immediately see a clock with each hour represented by a cluster of three notes. For example at twelve o'clock, we see a cluster of three notes (B, C, C-sharp) at the number one. The cluster of three notes can further be simplified if we identify each cluster as a point. We will now get the face of a clock with twelve hours, which reduces to the twelvefold cycle in the Western music scale. There is another peculiar thing about the diagram: Coltrane connects five repeating C notes within a five-sided star. What we have is cyclic geometry in the following sense: if you count the notes, you'll find sixty notes that repeat around the cycle. However, embedded in that cycle are twelve notes that generate five C notes within the sixty-note cycle—that's the star in the mandala. Coltrane's mandala, therefore, is a cycle embedded in a cycle.

When we identify all of the C notes in the five-sided star, we get the twelve-tone system of Western music. However, information is lost when we identify the five C notes as one C note in the mandala. By information I mean the geometry, or the five-sided object embedded in the sixty cycle. If I were to try to preserve the pentagon in our twelve-note cycle, we get a very interesting scale—the pentatonic scale. One is led to speculate, guided from Coltrane's statement to Amran, that "he was trying to do something like that [reduce complexity into something simple] in music, something that came from natural sources, the traditions of the blues and jazz." It is a fact that the pentatonic scale exists in cultures all over the world and dates back to China and Greece some

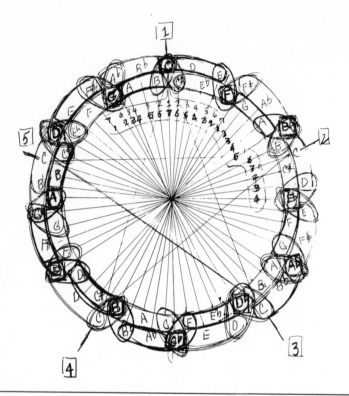

FIGURE 18.3. Coltrane's mandala, which reveals the fivefold cyclic geometry within the sixtyfold cycle. *Ayesha Lateef.*

twenty-five hundred years. The scale is widely used in Gregorian chants, Negro spirituals ("Nobody Knows the Trouble I've Seen"), Scottish music ("Auld Lang Syne"), Indian music, jazz standard ("I Got Rhythm," "Sweet Georgia Brown"), and rock ("Stairway to Heaven"). Coltrane was searching for what was universal in music, and the place to start was determining what aspect of music was universal across human cultures. He also said that he wanted to find music that came from natural sources. Well, the pentatonic scale can be generated from five perfect fifths. Recall that the perfect fifth is naturally generated as the second harmonic in the Fourier series, so this fulfills Coltrane's statement "that he was trying to do something that came from natural sources."

But the most compelling evidence is the fact that two of his most renowned pieces, *A Love Supreme* and *Interstellar Space,* are grounded in

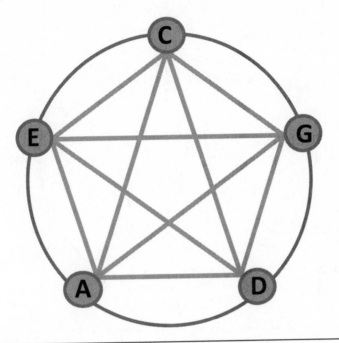

FIGURE 18.4. The fivefold symmetry of the C major pentatonic scale.

the pentatonic scale. Stacy Dillard, my friend and currently one of New York's most celebrated tenor saxophonists, said that the pentatonic scale is the skeleton of jazz improvisation. In other words, like Einstein's idea of invariance, the pentatonic scale is a basis from which complexity can unfold in a jazz improvisation. This does not say that the pentatonic scale is the unique basis, but it does raise questions as to why this relatively simple scale possesses such musical potentiality.

Coltrane's mandala also contains other beautiful relationships informed by cyclic geometry. And there are some resonances with Schoenberg and Messiaen who also used ideas in set theory in their compositions. Some important devices in jazz improvisation are the tritone substitution. All this really means is that in the passage from one chord to another, it is possible to replace the subsequent chord by another easier chord. We discussed that the II-V-I progression is one of the most common progressions in jazz and Western classical music. The

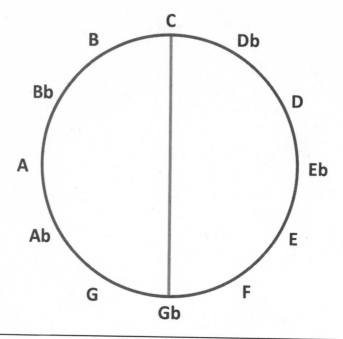

FIGURE 18.5. A mirror reflection relates a note to its tritone. For example, the tritone of C is F-sharp and the tritone of F-sharp is C.

tritone is nothing more than reflection symmetry in the twelve-tone cycle (see Figure 18.5). So in the key of C, the V is a G-dominant chord and its mirror image/tritone for G is D-flat dominant. Therefore, when we transit from G dominant to C, we can instead play D-flat dominant instead of G dominant. This is great because D-flat dominant is a half step from the II, which is D. Coltrane's sixty-cycle mandala also enjoys a reflection symmetry that has the tritone.

The cluster of three notes in the oval remarkably generates the mysterious all-interval tetrachord. For example, where we see the number one we begin with the C note and follow the next four notes in the cluster of ovals. We get C, C-sharp, E, F, and F-sharp, which is an all-interval tetrachord. It has been argued by Australian pianist Sean Wayland that the all-interval tetrachord can be used as a device to play through the chord changes of "Giant Steps."[2] There's more. Notice that Coltrane outlines a square within the three-note cluster. These notes are exactly the cycle of fifths, which generates the pentatonic scale. And finally Coltrane outlines one of the most widely used symmetric scales—the whole tone scale, which are the notes that occupy the inner ring and outer ring. Therefore the mandala is a stunning geometric creation by Coltrane that relates these important and general scales to each other, the same way space-time transformation relates length contraction to time dilation, electric fields to magnetic fields.

This book is not only about the analogy between music and cosmology but also about the importance of musical and improvisational thinking in doing physics. Theoretical physicists exemplify John Coltrane's approach to music. We use an arsenal of conceptual and mathematical tools that we practice through examples that were worked out by past masters, like Einstein and Feynman. Likewise, jazz musicians like Coltrane master their tradition throughout countless hours of practice. But for both the theoretical physicist and the jazz improviser, it is not enough to simply master the material of the past; discoveries must be made.

Humans are the only creatures that can discover advanced mathematics, and the only creatures that can create and formalize music. If the beauty and physics of the universe, and the beauty and physics of music, are linked, the links exist uniquely in human brains. Neuroscientists such as Rick Granger, György Buzsáki, and Ani Patel are still striving to understand how brains can perceive, learn and remember, and plan and predict. But even rats and dogs and bears can do all of those things. What, then, sets human brains apart? What makes us uniquely able to do what nonhuman brains cannot: appreciate music and understand mathematics? And to create new things under the sun: compose, improvise, discover new mathematical facts about the universe?

A few musicians, like Coltrane, have an uncanny ability to improvise, to find the hidden patterns and regularities underlying harmonic forms and to use those insights to generate brand-new kinds of melodic sequences. And a few scientists, like Einstein, can find regularities that have eluded even other great scientists—such as taking the Maxwell equations and reducing them into a single unifying formulation.

Perhaps we all have the immanent ability to do math like Einstein or to improvise like Coltrane. Perhaps their uniqueness lies in an ability to push those innate abilities far beyond the norm. Once the field of neuroscience has successfully captured the fundamentals of perception and thought, perhaps the next step can be to understand both what brains share and how brains differ and if new physics is needed—to understand what it is in Coltrane's brain and in Einstein's brain that lifted their thoughts to these insights and discoveries. Some current brain research is beginning to explore these questions: What is going on when we perceive the complexities of music? How do human brains process our environment so differently from other animals, giving us mathematics, musical improvisation, and language?[3]

To paraphrase the infamous pig from *Animal Farm*, apparently some human brains are more unique than others. Einstein and Coltrane showed us things that the rest of us had not discovered for ourselves. As we come to understand our brains, in general and in particular, perhaps neuroscience will begin to show us not just how musical form and physical form may be linked, but how it is that we, unique among physical beings, can come to see and understand these links.

Perhaps the answers to these questions require fundamental advances at the interface of physics, the arts, and neuroscience. The deep links between musical form and physical form may be unveiled by understanding how both kinds of knowledge—music and physics—arise together in human brains and nowhere else. After all, brains, regardless of how mysterious they are, are the most complex structures in the universe.

One of the fathers of calculus, Gottfried Leibniz, had the idea that the reducible element of the universe, the monad, had the capacity to contain the essence of the universe in it. It remains a mystery that human brains, which arise from and operate under the laws of physics, can come to understand the laws of physics. If one of the fundamental functions of the universe, as I've argued, is to improvise its structure, perhaps when Coltrane improvises, he is doing what the universe does, and what the universe did was to create a structure that would come to know the universe itself.

EPILOGUE

Behind any scientific discovery there are people and their stories. My journey, improvisational and eclectic, was guided by my physics and music mentors. For the last thirty years, I've been honored to learn secrets in the art of theoretical physics from Jim Gates. In 1969, he was a tall, slender young man, wearing bell-bottoms and sporting an Afro, who walked into MIT's infamous infinite corridor, pursuing his dream of becoming a physicist and an astronaut. The young James Gates Jr. would soon make friends with Ronald McNair, the astronaut who would suffer the tragedy of the *Challenger* space shuttle explosion. During those pioneering years, MIT admitted a group of black students of physics, including Shirley Jackson and Ronald McNair. This was only four years after President John F. Kennedy asked for legislation "giving all Americans the right to be served in facilities which are open to the public—hotels, restaurants, theaters, retail stores, and similar establishments," as well as "greater protection for the right to vote." These pioneering physicists paved the way for my generation of scientists to thrive.

Jim continued on at MIT and earned a PhD in theoretical physics, writing the first dissertation at MIT on supersymmetry and later a monstrous textbook, *A Thousand and One Lessons in Supersymmetry*. He joined Harvard's Society of Fellows, sharing an office with Michael

FIGURE EPI.1. *Left:* Jim Gates Jr. as a student at MIT. *Right:* Jim with Stephen Hawking.

Peskin (who, coincidentally, was my postdoctoral advisor), Edward Witten, and Warren Siegel. They were the crème de la crème in the field. Throughout the years, Jim and I mostly spoke about physics. While writing this book, I asked Jim about his time at Harvard. He said that he made friends with his office mates; they remain colleagues to this day. But back then, he was overwhelmed by the sheer brilliance of his colleagues Witten, Peskin, and Siegel.

While at Harvard, in 1977, Jim received an unexpected letter from Abdus Salam, a recent Nobel Prize winner for his discovery of the unification between the weak interaction and electromagnetism. The theorists of Jim's generation revered Salam as legendary, so just imagine everyone's reaction when they heard that Salam invited Jim to hang out with him and his group! At the time, Salam was working on supergravity, and Jim was a young pioneer on the subject, so it made sense for Jim to give a seminar. After winning the Nobel Prize, Salam launched the International Center for Theoretical Physics (ICTP), whose mission was to "develop high-level scientific programmes, keeping in mind the needs of developing countries, and provide an international forum of scientific contact for scientists from all countries." There, for the first time in Jim's career, he saw physicists from all corners of the globe— Africa, China, Europe, and the Middle East. He realized that physics truly was a global enterprise, not restricted to Europe and America, a common misconception promulgated by mainstream media.

After the seminar, Salam took Jim out for lunch. Jim was full of questions and ideas to share with the guru. Then, out of the blue, Salam says to Jim: "One day, when your people do physics, it will be like jazz."

What a great compliment, an affirmation and acknowledgment of the improvisational, inclusive, cultural, and intellectual contributions of this music called jazz. Embodied in Salam's statement is how the genius of a people who have created a musical culture as dynamic in sound and rich in metaphor as jazz can contribute to the very enterprise of physics.

Learning how to play this music we call jazz is a lifelong process. The music has evolved over the nearly one hundred years of its recorded existence into an intellectually and artistically demanding system. For example, bebop jazz was 99.9 percent developed outside the academy—during the rehearsals and the jam and cutting sessions of the forties—by such greats as Charlie Parker, Dizzy Gillespie, Bud Powell, Max Roach, and Thelonious Monk, to name a few. These artists lived during a time when Jim Crow was the de facto law of the land and lynchings still occurred. So how did this music get to be so great, to the point where it is called America's classical music, arguably the most representative art form created in the United States of America? In the classic *The Hero and the Blues*, one of Wynton Marsalis's mentors, Albert Murray, contends that "antagonistic cooperation" is behind the greatness in the jazz tradition. In a conversation with Marsalis, printed in the forthcoming University of Minnesota book *Murray Talks Music*, Murray says:

Let us bring another concept in here from my little book *The Hero and the Blues* and that is the concept of antagonistic cooperation. Sort of a contradiction in terms but it adds up to a mnemonic device, which is very useful. If you don't have adequate opposition, you don't develop. To be a great champion you have to have great contenders. To be a great hero you have to have dragons to kill.

But even while there was fierce competition, which enabled growth, jazz musicians were also inclusive. Anyone was able to get on the bandstand and express him- or herself.[1] If you were good, you were called back to play. The tradition was also embracing. I remember times when I was the least skilled player and yet would be invited to play with greats like Will Calhoun, Marc Cary, and John Benitez. These were Grammy award–winning musicians. Even though my solos weren't the most polished, serendipitous moments—where an interesting phrase, sometimes a weird one, would surface—Will would take note of those ideas and would sometimes incorporate them into another song.

So when I heard Salam's story, that one day physics could look like jazz, I interpreted it as one day when the physics community, like jazz, includes contributions from all groups of people, regardless of creed, it will reach new heights, allowing us to solve problems once thought to be impossible. My journey to reconcile jazz with physics serves as a living example of how a small group of physicists, in the spirit of the jazz tradition, embraced me and allowed me to improvise physics with them, while challenging me to go beyond my limits.

ACKNOWLEDGMENTS

Special thanks to my editor TJ Kelleher, the magician, and to Lara Heimert. Thanks to the entire Basic Books and Perseus staff—Helene Barthelemy, Sandra Beris, Cassie Nelson, Liz Tzetzo—for making this book come into reality. Eternal thanks to Dagny Kimberly Yousuf for helping me get the original manuscript in shape, and all of the inspiration and support from day one of writing this book. Thanks to Max Brockman and Brockman Inc. staff for helping to make the book a reality.

I also thank my friends, family, and colleagues who supplied hours of inspiration, feedback, ideas, and encouragement: Rome Alexander, Steven Beckerman, Robert Caldwell, Will Calhoun, Steve Canon, Michael Casey, KC Cole, Ornette Coleman, Diego Cortez, François Dorias, Brian Eno, Everard Findlay, Edward Frenkel, Margaret Geller, Indradeep Ghosh, Melvin Gibbs, Marcelo Gleiser, Rebecca Goldstein, Mark Gould, Rick Granger, Daniel Grin, Sam Heydt, Chris Hull, Chris Isham, Beth Jacobs, Clifford Johnson, Brian Keating, Jaron Lanier, Yusef Lateef, Harry Lennix, Arto Lindsa, Joao Magueijo, Brandon Ogbunu, Steve Pinker, Sanjaye Ramgoolam, Erin Rioux, Tristan Smith, Lee Smolin, David Spergel, Greg Tate, Greg Thomas, Spencer Topel, Gary Weber, and Eric Weinstein. My gratitude to Salvador Almagro-Moreno for his brilliant custom diagrams throughout this book.

NOTES

INTRODUCTION

1. Yusef Lateef, *Repository of Scales and Melodic Patterns* (Amherst, MA: Fana Music, 1981).

2. Yusef Lateef, *The Gentle Giant: The Autobiography of Yusef Lateef*, with Herb Boyd (Irvington, NJ: Morton Books, 2006).

CHAPTER 1

1. It has been argued that birds also make music for their listening pleasure.

2. Other authors enlarge the number of "dimensions" to music. I refer the reader to the wonderful book *This Is Your Brain on Music: The Science of a Human Obsession,* by Daniel J. Levitin (New York: Plume, 2007), for a more complete set of twelve dimensions of music perception.

3. According to Benoit Mandelbrot, one of the pioneers of fractal geometry, "assembled artfully, the combination is supposed to fit together harmoniously—somewhat like a coastline." Interestingly, in his book *Fractals: Form, Chance, and Dimension* (San Francisco: W. H. Freeman, 1977), Mandelbrot argued that galaxies organize themselves into a fractal structure. Soon after, astrophysicist Luciano Pietronero found that systems of galaxies did indeed possess a fractal structure, although this claim remains debatable.

4. Malcolm Brown, "J. S. Bach + Fractals = New Music," science section, *New York Times*, April 16, 1991, www.nytimes.com/1991/04/16/science/j-s -bach-fractals-new-music.html?pagewanted=1, accessed November 28, 2015.

5. Charles W. Misner, Kips S. Thorne, and John Archibald Wheeler, *Gravitation* (San Francisco: W. H. Freeman, 1973).

CHAPTER 2

1. This model is otherwise known as the Heisenberg XXX model. I thank Edward Frenkel for pointing this out to me.

2. Magnetic material exhibit a Hysteresis curve, which is a "memory" of how magnetized a magnet gets when an external magnetic field is applied.

3. Carl Clements, "John Coltrane and the Integration of Indian Concepts in Jazz Improvisation," *Jazz Research Journal* 2, no. 2 (2008): 155–175.

CHAPTER 3

1. In jazz a section of a song tends to repeat itself while a soloist is improvising. Right before the section ends, there is a harmonic movement (a turnaround) that enables the song to transition back to the beginning of the song.

2. Personal communication with Margaret Geller, October 2015.

3. Margaret J. Geller and John P. Huchra, "Mapping the Universe," *Science* 246, no. 4932 (November 17, 1989): 897–903, doi:10.1126/science.246 .4932.897, PMID 17812575, retrieved May 3, 2011.

4. We will see later that string theory resolves this infinity.

CHAPTER 4

1. Michio Kaku, "The Universe Is a Symphony of Strings," Big Think, http:// bigthink.com/dr-kakus-universe/the-universe-is-a-symphony-of-vibrating -strings, accessed November 28, 2015.

2. In physics, especially quantum mechanics, many issues of interpretation of the theory transpire. Some physicists take the attitude that there is no room for subjective interpretation and adopt the attitude that one should just calculate results, hence the phrase "shut up and calculate." It is thought that Dirac and Feynman originated this term; credit is also given to David Mermin, a solid-state physicist.

3. Steven Weinberg, *Dreams of a Final Theory: The Scientist's Search for the Ultimate Laws of Nature* (New York: Pantheon, 1993), 130. Dirac, the famous proponent of beauty, later realized that when Einstein's postulate was incorporated into the equation describing the motion of the electron, a hidden

symmetry revealed itself. Changing a sign from positive to negative implied the physical equivalent of changing the charge of the electron. To his surprise, this resulted in physics consistent with quantum mechanics, which implied the existence of a previously unknown particle. A year later this antielectron, or positron, with equal mass and opposite charge to the electron, was discovered and won Dirac the Nobel Prize.

4. The conceptual and mathematical idea of spontaneous symmetry breaking will be discussed in Chapter 11, using ideas in music theory.

5. David Demsey, "Chromatic Third Relations in the Music of John Coltrane," *Annual Review of Jazz Studies* 5 (1991): 145–180; Demsey, "Earthly Origins of Coltrane's Thirds Cycles," *Downbeat* 62, no. 7 (1995): 63.

CHAPTER 5

1. Marcelo Gleiser, *The Dancing Universe: From Creation Myths to the Big Bang* (Lebanon, NH: University Press of New England, 1997).

2. Jamie James, *The Music of the Spheres* (New York: Grove Press, 1933), 64: "Musica instrumentalis is harmonious because it reflects the perfection of the cosmos in the world of ideal forms; an octave sounds harmonious to human ears because the rhythms of the music are in concord with our own internal rhythms . . . the musica humana."

3. Gleiser, *The Dancing Universe*.

4. Ibid.

5. Willie Ruff and John Rodgers, *The Harmony of the World: A Realization for the Ear of Johannes Kepler's Astronomical Data from Harmonices Mundi 1619*, Kepler Label, August 3, 2011, compact disc.

6. Johannes Kepler, *The Secret of the Universe: Mysterium Cosmographicum*, trans. A. M. Duncan (New York: Abaris Books, 1981).

CHAPTER 6

1. From "Generative Music: Evolving Metaphors, in My Opinion, Is What Artists Do," a talk Brian Eno delivered in San Francisco, June 8, 1996.

CHAPTER 7

1. See http://americansongwriter.com/2008/01/ornette-coleman-the-language-of-sound/.

2. AAJ Staff, "A Fireside Chat with Marc Ribot," *All About Jazz*, February 21, 2004, www.allaboutjazz.com/a-fireside-chat-with-marc-ribot-marc-ribot-by-aaj-staff.php, accessed November 28, 2015.

CHAPTER 8

1. There is a subtlety in that the second derivative of sin(t) is −sin(t).

2. Larry Hardesty, "The Faster-Than-Fast Fourier Transform," Phys.Org, January 18, 2012, http://phys.org/news/2012–01-faster-than-fast-fourier.html, accessed November 28, 2015.

3. The cosine function is a shifted sine wave and so is just as fundamental as the sine wave. Technically, it would appear here, but because it is a derivative of the sine function, this simplified equation satisfies our purposes.

CHAPTER 9

1. See jpeg of icosahedraon at www.twiv.tv/wp-content/uploads/2009/07/icosahedral-symmetry-1024x775.jpg.

2. P. W. Anderson, "More Is Different," *Science* 177, no. 4047 (1972): 393–396.

CHAPTER 10

1. To be precise, Earth is in motion, rotating around the sun, but this amount of rotational acceleration is too small to be felt on human scales.

CHAPTER 11

1. "Interpreting the 'Song' of a Distant Black Hole," Goddard Space Flight Center, NASA, November 17, 2003, www.nasa.gov/centers/goddard/universe/black_hole_sound.html, accessed November 28, 2015.

CHAPTER 12

1. "Interpreting the 'Song' of a Distant Black Hole," Goddard Space Flight Center, NASA, November 17, 2003, www.nasa.gov/centers/goddard/universe/black_hole_sound.html, accessed November 28, 2015.

2. Rashid Sunyaev and Ya Zel'dovich simultaneously arrived at the same conclusion.

3. There are exceptions where we can perceive a pitch even though the fundamental is missing from the physical sound spectrum.

4. John Cage, "Forerunners of Modern Music," in *Silence: Lectures and Writings* (Middletown, CT: Wesleyan University Press, 1961), 62.

CHAPTER 13

1. An arpeggio is a chord in which the individual notes are played in either an ascending or descending order.

2. Transcription in the context of jazz repertoire is the act of annotating a solo of a jazz recording. A student of jazz will analyze the voicing of the solo in relation to the chord changes and commit parts of the solo to memory in order to develop a more lyrical vocabulary.

3. Devin Leonard, "Mark Coltrane Escapes the Shadow of John Coltrane," *Observer,* June 26, 2009, http://observer.com/2009/06/mark-turner-escapes-the -shadow-of-john-coltrane/, accessed November 9, 2015.

4. "Warne Marsh & Lennie Tristano Discuss Improvisation," YouTube video, 1:26, uploaded March 6, 2011, www.youtube.com/watch?v=YqSdXxw bfM0.

5. Roger Highfield and Paul Carter, *The Private Lives of Albert Einstein* (New York: St. Martin's Griffin, 1995).

CHAPTER 14

1. www.azquotes.com/author/9502-Wynton_Marsalis/tag/jazz, accessed November 18, 2015.

2. Gunther Schuller, "Sonny Rollins and the Challenge of Thematic Improvisation," *The Jazz Review,* November 1958, http://jazzstudiesonline.org/files /jso/resources/pdf/SonnyRollinsAndChallengeOfThematicImprov.pdf, accessed November 9, 2015.

3. Jonah, "The Graphene Electro-Optic Modulator," The Physics Mill, May 25, 2014, www.thephysicsmill.com/page/5/, accessed November 28, 2015.

CHAPTER 15

1. Rhiannon Gwyn et al., "Magnetic Fields from Heterotic Cosmic Strings," *Physics in Canada* 64, no. 3 (Summer 2008): 132–133. One approach to addressing the origin of primordial magnetic fields may come from heterotic

string theory, which was presented in a paper developed by Stephon Alexander and his collaborators. The heterotic string can act like a wire, which carries an electric charge, generating a magnetic field. If the early universe is filled with a homogenous network of these heterotic "cosmic strings" then a suitable amount of primordial galactic magnetic fields can be generated.

CHAPTER 16

1. Allan Kozinn, "John Cage, 79, a Minimalist Enchanted with Sound, Dies," *New York Times*, August 13, 1992, www.nytimes.com/1992/08/13/us /john-cage-79-a-minimalist-enchanted-with-sound-dies.html, accessed November 28, 2015.

2. Spin zero fields are described by scalar functions, such as F (x). In contrast, spin one fields, such as electromagnetic fields, are described by vector functions. The vector indices in a vector function give information about the polarization of the field.

3. In quantum field theory, fields typically self-interact. This usually corresponds to the ability of a field to create many particles starting from one particle. Self-interactions also characterize the potential energy stored in a quantum field.

4. Lisa Randall, *Warped Passages: Unravelling the Mysteries of the Universe's Hidden Dimensions* (New York: HarperCollins, 2005).

CHAPTER 17

1. I and, independently, the Greek string theorist Elias Kiritsis figured out a clever way of realizing João Magueijo's mechanism in string theory. The idea again relied on D-branes. Black holes can also live in a five-dimensional universe. At the center of our galaxy is Sagittarius A-star, a supermassive black hole partnered with the stars orbiting around it. Suppose a D3-brane universe is orbiting around a five-dimensional black hole. What Kiritsis and I discovered was that the speed of light in the brane universe would change with the distance the brane is from the black hole.

2. Personal communication with Spencer Topel.

3. Nima Arkani-Hamed et al., "Ghost Condensation and a Consistent Infrared Modification of Gravity," *Journal of High Energy Physics* 405 (2004): 74.

4. If the volume of the universe gets larger, the entropy will increase, which implies a longer cycle.

CHAPTER 18

1. Ben Ratliff, *Coltrane: The Story of a Sound* (New York: Farrar, Straus and Giroux, 2007).

2. See the enlightening video on using the all-interval tetrachord for "Giant Steps" at www.youtube.com/watch?v=sQGWAnYd7Iw.

3. Y. S. Lee et al., "Multivariate Sensitivity to Voice during Auditory Categorization," *Journal of Neurophysiology* 114, no. 3 (2015): 1819–1826; Y. S. Lee et al., "Melody Revisited: Influence of Melodic Gestalt on the Encoding of Relational Pitch Information," *Psychonomic Bulletin and Review* 22, no. 1 (February 2015): 163–169; Richard Granger, "How Brains Are Built: Principles of Computational Neuroscience," *Cerebrum*, The Dana Foundation, January 31, 2011, http://dana.org/news/cerebrum/detail.aspx?id=30356, accessed November 28, 2015.

EPILOGUE

1. I thank Eric Weinstein for inspiring me on this important fact.

INDEX

acceleration, 104

acoustic peaks, 152–153

Afrika Bambaataa (hip-hop artist), 24

Aion: Researches into the Phenomenology of Self (Jung), 60

Albrecht, Andy, 203

Alexander, Eric, 173

Alpher, Ralph, 129

Ambient 1: Music for Airports (Eno), 89

ambient music, 89

Amran, David, 64, 218–219

analogical reasoning, 2

analogical thinking, 3, 83

analogies

    jazz music and cosmology, 100

    Kepler on, 1, 83

    in music, 40, 69

    music and science, 98

    power of, 6–7, 207

    in science, 39–40, 69

Anderson, Phil, 121

anisotropy, 132, 132 (fig.), 133, 146

"The Anthropic Landscape of String Theory" (Susskind), 196

anthropic principle, 196

antigravity, 5

antiparticles, 167, 168 (fig.)

Aristotle, 73

Arkani-Hamed, Nima, 210

Armstrong, Louis, 96

AS220 (Providence, Rhode Island), 41, 42

astrologers, 44–45

astronomy, 69, 76

astrophysicist, 77

"Auld Lang Syne" (song), 224

autophysiopsychic music, 4

Bach, Johann Sebastian, 18, 65, 157

Bambrick, Miss (teacher), 20

Bardeen, John, 28

Barrow, John, 211

baryogenesis, 182–183, 185–186, 188

baryon current, 184–185

battle rap, 20, 22

beats, 16

Beethoven, Ludwig van, 16, 98

Benitez, John, 232

bilateral symmetry, 120

biophysics, 117

black bodies, 164

black holes, 139–144

"Blue 7" (song), 173

blues, music, 17

blues scale (music), 172, 173 (fig.)

blues structure, 17 (fig.)

Bohr, Neils, 167

boson fields, 181

bosons, 58
Bowie, David, 88
Brahe, Tycho, 76–77
Brandenberger, Robert
blending physics and music, 69
cosmic inflation, 123
fundamental constants, 204
as mentor, 42–44, 62, 84
neural network, 48–50
photo of, 43 (photo)
preheat phase, 193
string theory, 52–53
branes, 86–87, 197–199, 198 (fig.)
A Brief History of Time (Hawking), 31,
59
Brindle, Hill, 18–19
Broglie, Louis de, 164–165, 166–167
broken quantum symmetries, 7
broken symmetries, 122–123
Brown University, 41, 53, 119
bubble universe, 195 (fig.), 196, 211
Buzsaki, György, 227

Cage, John, 157, 189
calculus, 102, 107
Calhoun, Will, 232
calypso music, 12
Cary, Marc, 232
Casey, Michael, 177
causal set theory, 86
Challenger space shuttle, 229
Chandrasekhar, Subrahmanyan, 139
change, mathematics of, 102
chords, 16–17, 65–66
City College of New York, 24
Clarence 13X (Five Percenter Nation
leader), 21
classic mainstream jazz, 95–97
classical mechanics, 102
classical particles, 176 (fig.)
classical Western music, 16
CMB. *See* cosmic microwave
background
CMB anisotropy, 132, 132 (fig.), 133

COBE (Cosmic Background Explorer)
satellite, 135, 146, 193
Coldplay (English rock band), 88
Coleman, George, 173
Coleman, Ornette, 8, 15, 41, 95,
97–100
The Collective (jazz fusion band), 42
Coltrane, John
ancient philosophy study, 209
Cosmic Sound, 5, 218
death of, 5
diagram given to Lateef, 3–4, 5 (fig.)
Einstein as idol, 217–218
fusing music traditions, 40
Giant Steps (album), 24, 64, 160,
217
"Giant Steps" (song), 50, 209,
211, 225
Interstellar Space, 3, 5, 6, 218, 224
"Jupiter," 218
last recorded albums, 5
"A Love Supreme," 160, 224
musical talent of, 160
"My Favorite Things," 40
relativity and music, 222
search for meaning of the cosmos,
217–218, 221
Stellar Regions, 218
symmetric scale, 64–65
Coltrane, Ravi, 218
Coltrane's mandala
gift to Lateef, 5, 5 (fig.)
symmetry and, 65
thought experiment using, 217,
222–223, 224 (fig.), 225,
227 (fig.)
Columbia University's Institute
for String Cosmology and
Astrophysics (ISCAP), 93
comets, 77
communicating ideas, 2
complementarity, wave-particle, 167
complexity, 85
conservation of energy, 22–23

consonant notes, 80
consonant tone, 71, 72
constant force, 105–108, 106 (fig.)
Contemporary Population I stars, 155
Cooper, Leon
    analogies, 100, 207
    analogy breakdown, 153
    as mentor, 8, 84
    neuroscience, study in, 33
    photo of, 30 (photo)
    as professor, 27
    superconductivity theory, 29–30
Cooper group, 32
Cooper pair (superconductivity
        theory), 29–30
Copernican Revolution, 76
Copernicus, Nicholas, 75–76
cosine waves, 111
Cosmic Background Explorer (COBE)
        satellite, 135, 146, 193
cosmic bounce, 209–210
cosmic expansion
    antigravity and, 5
    cosmologists and, 48
    Einstein and, 127–129
    inflation and, 191–201
cosmic horizon, 144
cosmic inflation. See inflation theory
cosmic microwave background (CMB),
        130–131, 134 (fig.), 146,
        181–182, 189, 207
Cosmic Mystery (Kepler), 78
Cosmic Sound (Coltrane), 5, 218
cosmological constant, 127
cosmological principle, 136
cosmology, 44, 45, 47–48
cosmos, harmonic nature of, 71
Coulomb force, 129
coupling constants, 190, 195–196,
        199, 201, 211–213
Crab Nebula, 74, 75 (fig.)
creation myths, 69
Crook, Hal, 41, 42, 173
cube, as platonic solid, 73

Curtis, Heber, 45
curved space, 126–127
cyclic geometry, 222, 223 (fig.), 225
cyclic universe, 208–212, 209 (fig.),
        212 (fig.), 216

D1-branes, 200
D3-branes, 198 (fig.), 199, 200
D5-branes, 199, 200
dark energy, 94
dark matter, 153, 154–155, 190
data mining, 39
D-branes, 197, 199–201
De Broglie, Louis, 164–165, 166–167
De La Soul (hip-hop trio), 24
Debussy, Claude, 63
DeWitt Clinton High School (Bronx,
        NY), 20
Di Dario, Mrs. (teacher), 12, 42
Dickie, Robert, 130
Dickie radiometer, 130
differential equations, 107, 137, 138
differential geometry, 42, 126
differential microwave radiometer, 135
Dillard, Stacy, 225
Dirac, Paul, 51, 54, 57, 62, 72, 167
Discreet Music (Eno), 91
divine geometry, 74, 101
DNA structure, 47
dodecahedron (twelve pentagons), 73
double helix structure of DNA, 47
Dowker, Faye, 86
Dreams of a Final Theory (Weinberg),
        54

Earth
    as center of universe, 72–73
    Kepler's model, 78
Einstein, Albert
    Coltrane's idol, 217–218
    cosmological constant, 127
    equivalence principle, 126
    four-dimensional space-time, 55
    gedankenexperiments, 4, 202

Einstein, Albert *(continued)*
    general relativity theory, 4, 126,
        127, 208, 209
    gravity, 24, 126
    interdisciplinary focus, 3
    on light, 163–164
    mind of God, 52
    museum exhibit, 13–14
    music and, 1, 4–5
    photoelectric effect, 164
    on protons, 163–164
    relation E=hf, 164, 183
    space-time structure, 55, 86,
        125, 127
    space-time symmetry, 221–222
    special relativity, 65, 120, 204,
        219–220, 222
    speed of light, 204
    symmetric scale and, 64–65
    on the universe, 216
    universe expansion hypothesis,
        127–129
Einstein, Elsa, 4
electricity, 29
electromagnet radiation, 13
electromagnetic theory, 163, 204
electromagnetics and weak nuclear
        force, 54–55
electroweak force, 54
elegant equations, 55
elementary particle physics, 56, 122
elements, 73
eleven-dimensional supergravity, 57
elliptical orbits, 81
emergent phenomenon, 33
energy
    conservation of, 22–23
    types of, 34
Eno, Brian
    ambient music, 89
    complexity, 85
    frequency modulation synthesis,
        113 (fig.)
    generative music, 90–91
    as mentor, 8
    music, 87–88, 98
    sound cosmologist, 88
equivalence principle, 126
Eric B is President (Rakim), 21
ether, element, 73
event horizon, 144

false vacuum energy, 195 (fig.)
Farley, Ruby, 11–13
Feder, Daniel, 21
Fergurson, Harvey, 24, 25
fermions, 58
fermions fields, 181
ferromagnet, 38, 191
Feynman, Richard, 27, 175–176
Feynman diagrams, 27–29, 28 (fig.),
        34, 175
Feynman path integral, 175–178
fiber bundle theory, 54
field strength tensor, 221
Fifth Symphony (Beethoven), 16, 98
fine-tuning problem
    cosmological constant, 190
    cyclic universe, 208–212, 212 (fig.),
        216
    inflation theory and, 194–195,
        197, 200
    quantum field theory, 194
Finkelstein, David, 139–140
Five Percenter Nation, 20–21
force, 103, 104–105
form, defined, 16
four-dimensional current, 221
four-dimensional space-time, 55,
        219 (fig.)
Fourier idea, 111, 112 (fig.), 147,
        166, 190
Fourier transform, 90, 101, 112–113,
        147 (fig.)
fractal structures, 18
free jazz, 15, 160
free-falling masses, 102
free-style battle rap, 20

frequency, 165–166
frequency, fundamental, 148, 152
frequency modulation synthesis, 113 (fig.)
The Fringe (jazz band), 41
From Atom to Archetype (Pauli & Jung), 61
fruit fly experiments, 94–95
fundamental frequency, 148, 152

galaxies, 45–49
Galileans (Jupiter moons), 76
Galilei, Galileo, 45, 76, 101
Gamow, George, 129
Gates, James, Jr., 229
Gates, Jim, 117, 229–231
gauge potential, 220
gedankenexperiments, 4, 202
Geller, Margaret, 45–47, 47 (photo), 136, 152
Geller, Seymour, 45, 152
general relativity theory, 4, 126, 127, 208, 209
generative music, 90–91
geocentrism, 76, 77
geometrical representation, 63
geometry
    cyclic, 222, 223 (fig.), 225
    differential, 42, 126
    divine, 74, 101
    noncommutative, 61
Gershwin, George, 66
ghost field, 210
Giant Steps (Coltrane album), 24, 64, 160, 217
"Giant Steps" (Coltrane song), 50, 209, 211, 226
Gilbert, Wally, 119
Gillespie, Dizzy, 231
Gleiser, Marcelo, 212
Granger, Rick, 227
gravitation, 55
Gravitation (Einstein), 24
gravitational instability, 153 (fig.)

gravitational wave, 187–188
gravity
    Einstein on, 24, 126
    Newton on, 82–83, 102
    planetary motions and, 79
    quantum mechanics and, 49, 52
    sound waves and, 151
Great Wall (galaxies), 46, 46 (fig.)
Greene, Brian, 93
Gregorian chants, 224
Gregory, Fredrick, 19
Guinan, Ed, 156
Guralnik, Gerry, 118
Guth, Alan, 123, 133–135, 191, 192, 193

Handler, Mrs. (teacher), 13
harmolodics, 96, 97
harmony, 16, 71
The Harmony of the World (Kepler), 81, 82 (fig.)
Harrison, Donald, 171–172, 176
Harvard Medical School, 119
Hawking, Stephen, 31, 59, 86
Hawkins, Coleman, 97
Heisenberg, Werner, 31, 162, 165, 177
heliocentrism, 76
helioseismology, 156
helix, as quasi-periodic structure, 118, 118 (fig.)
Here Comes Now (Alexander & Rioux), 98, 99 (fig.)
Herman, Robert, 129
The Hero and the Blues (Murray), 231–232
Higgs boson particle, 7, 31
hip-hop music, 14, 24–25
Hogle, Jim, 120, 122
Hopfield, John, 38
Hopfield model of neuroscience, 33, 38–39, 40
horizon (invisible spherical surface), 140, 144
horizon, sonic, 142 (fig.), 143–144

horizon problem, 131
Hubble, Edwin, 45, 127, 129
Hubble space telescope, 135–136
Huchra, John, 46 (fig.), 46–47,
    136, 152
Hull, Chris, 197
human and universal origins, 1–2
hypothesizing and testing, 80

"I Can't Help Falling in Love" (song),
    12, 71
"I Got Rhythm" (song), 224
icosahedral symmetry, 120, 121 (fig.)
icosahedron (twenty triangles), 73
ICTP (International Center for
    Theoretical Physics), 230–231
Imperial College (London), 53, 85
improvisation
    characteristics of music and physics,
        5, 62, 161
    Marsalis on, 172–173
    music and, 14–15
    Ornette's lesson, 95–97
improvisational physics, 7
improvisational thinking, 2
Indian music, 224
inertia, 102, 104
inflation field, 135, 187–188
"Inflation from D-Anti-D brane
    Inflation" (Alexander), 201
inflation theory
    baryogenesis and, 182, 186, 188
    cosmic expansion and, 191–201
    early universe and, 190, 195 (fig.)
    fine-tuning problem, 194–195,
        197, 200
    Guth's theory of, 123, 133
    loudness of primordial sound, 194
    noise and, 190–194
    Occam's razor model of, 202
    quantum fluctuations, 193
    string theory and, 53, 196–201
    stringy inflation models, 201
    uniformity of CMB, 134 (fig.)
    work of inflation, 135

Institut Henri Poincaré (Paris), 197
Institute for String Cosmology and
    Astrophysics (ISCAP), 93
interaction energy, 34
International Center for Theoretical
    Physics (ICTP), 230–231
Interstellar Space, 215–228, 227 (fig.)
Interstellar Space (Coltrane), 3, 5, 6,
    218, 224
inverse Fourier transform, 112–113
ISCAP (Institute for String Cosmology
    and Astrophysics), 93
Isham, Chris, 59–60, 62, 69, 84,
    87, 185
Ising, Ernst, 33
Ising model of magnetism, 33–39,
    37 (fig.), 118, 137, 181, 191
isometry, 63

Jackson, Shirley, 229
jazz music, 231–232
jazz vocabulary, 172–173
Jazzy Jay (hip-hop artist), 24–25
Jevicki, Antal, 53
John Philip Sousa Junior High (Bronx,
    NY), 15, 18–20
Jung, Carl, 60
Jungle Brothers (hip-hop artist), 24
Jupiter, 76, 78, 82
"Jupiter" (Coltrane), 218

Kaku, Michio, 51–52
Kalkkinen, Jussi, 54
Kamerlingh Onnes, Heike, 29
Kaplan, Daniel, 22–23, 25, 103
Kennedy, John F., 229
Kepler, Johannes
    analogous thought, 1, 83
    Brahe and, 76–82
    Cosmic Mystery, 78
    earth model, 78
    elliptical motion of planets, 55
    The Harmony of the World, 81,
        82 (fig.)
    laws of planetary motion, 81

music and, 206
music to explain the universe, 50
musical notes of planets, 81–82,
    82 (fig.)
The Mysterium Cosmographicum, 78,
    79 (fig.)
planetary motion, 101
Platonic solid model, 79 (fig.)
second law, 81 (fig.)
string theory and, 6
keys (musical), 16–17
Kimberly, Dagny, 211
King, B. B., 17
Kleban, Matt, 196

Lanier, Jaron, 94–95, 98
Laplace, Pierre-Simon, 162
Large Hadron Collider, 7
Lateef, Yusef, 3–4, 5 (fig.), 222
law of gravitation, 55, 83
law of receding galaxies, 129
laws of motion, 81, 102–104, 114
laws of planetary motion, 81
Leibniz, Gottfried, 11, 22, 107, 215
The Life of the Cosmos (Smolin), 196
Ligetti, György, 18, 157
light
    impact on gravity, 151
    motion of, 168 (fig.)
    speed of, 151, 203–206
light radiation, 130–131, 131 (fig.),
    202
light waves, 153
loop quantum gravity, 86
Lord Kitchener, 12
Lorentz symmetry, 204
A Love Supreme (Coltrane), 160,
    224
lunar eclipse, 77
Lux (Eno), 91

magnetic fields, 179, 180 (fig.)
magnetic resonance imaging (MRI),
    30–31
magnetic train levitation, 31

magnetism, Ising model of, 33–39,
    37 (fig.), 118, 137, 181, 191
Magueijo, Joäo, 61, 203–206,
    205 (photo), 211
Marciano, Antonino, 202
Mars, 77, 79, 82
Marsalis, Wynton, 172, 231
Marsh, Warne, 160–161
Martino, Pat, 64, 65
*Mathematical Principles of Natural
    Philosophy* (Newton), 82
mathematical symmetry, 138
mathematics, 34, 39, 70–73
matrix mechanics, 31
matter, space-time and, 126
Maxwell, James Clerk, 180
Maxwell's equations, 55, 220 (fig.),
    220–221
McNair, Ronald, 229
Meissner effect, 31
melody, 16
memory, 33, 38
Mercury, 78, 127
Messiaen, Oliver, 225
meter, 16
Mighty Sparrow (calypso musician), 12
mind of God, Einstein writing on, 52
Minkowski space, 219 (fig.)
moiré patterns, 90–91
monad, 228
Monk, Thelonious, 231
moons of Jupiter, 76
"More Is Different" (Anderson), 121
motif, 97–98
motion, laws of, 81, 102–103, 114
Mozart, Wolfgang Amadeus, 4, 8
MRI (magnetic resonance imaging),
    30–31
M-theory, 197–199
Murray, Albert, 231–232
Murray Talks Music (Murray), 231–232
music
    astronomy and, 69
    autophysiopsychic, 4
    battle rap, 20, 22

music *(continued)*
　blues, 17
　blues scale, 172, 173 (fig.)
　calypso, 12
　characteristics of, 16–18
　classical Western, 16
　connections between universe and, 217–218
　engaging the unconscious and, 4, 60
　free-style battle rap, 20
　hip-hop, 14, 24
　jazz, 231–232
　jazz vocabulary, 172–173
　rules of, 96
　sense of purpose, 216
　soca music, 12
　spontaneity in, 161
　stars and, 156–157
　straight ahead jazz, 95–97
　symmetric scale, 63–64, 64 (fig.)
　universe and, 207–208
　Western scale, 70–71, 223
music theory, 3
musical cosmos, 145
musical instruments, 114, 148–149, 150 (fig.), 152
musical notes of planets, 82 (fig.)
"My Favorite Things" (Coltrane), 40
"My Melody" (Rakim), 22
The Mysterium Cosmographicum (Kepler), 78, 79 (fig.)
myth, 1–2

Native Tongue (hip-hop artist), 24
nebulae, 45
negative energy particles, 167
Negro spirituals, 224
Neptune, 77–78, 156
neural circuitry, 38
neural network, 33, 38–39, 49, 94
neurons, 38–39, 40
neuroscience, 33, 227–228
neutrino, 61
neutron stars, 139

Newton, Isaac
　calculus, 102, 107
　constant force, 105–108, 106 (fig.)
　force and velocity, 103–104
　gravitation, 55, 82, 106
　inertia, 102, 104
　laws of motion, 102–104, 114
　mathematical predecessors and, 101–102
　nonconstant force, 108–110, 109 (fig.)
　strings and waves, 6
"Night and Day" (Porter), 17
"Nobody Knows the Trouble I've Seen" (song), 224
noise
　as foundation for structure, 190
　Fourier idea and, 190
　inflation theory and, 190–194
　unwanted, 189–190
　See also sound
noncommutative geometry, 61
nonconstant force, 108–110, 109 (fig.)
nonliving matter, 118
nothingness state, 195

observational astronomy, 76
Occam's razor model of inflation, 202
Ocean Coffee (Providence, Rhode Island), 44
octahedrons (eight triangles), 73
octaves, musical, 71
Ogbunu, Brandon, 120
Oppenheimer, Robert, 139

parabola sketch of constant force, 106 (fig.), 107
parental chords, 64
Parker, Charlie, 96, 231
particle physics, 56, 122
Partridge, Bruce, 132–135
Patel, Ani, 227
Pauli, Wolfgang, 60–61, 62
p-branes, 86–87

Peebles, Jim, 130, 145, 151, 152
Penrose, Roger, 59
pentatonic scale, 223–225, 224 (fig.)
Penzias, Arno, 5, 130, 189
Perry, Sacha, 41
Perseus cluster, sound waves, 143 (fig.)
Peskin, Michael, 186–188, 187
    (photo), 229–230
photoelectric effect, 164
"Photographing the Wavefunction of
    the Universe" (Magueijo), 205
photon's energy and frequency,
    164–165
phrases (musical), 16
pitch, 151
Piteo, Paul, 15, 20
Planck, Max, 163, 164, 165
Planck space observatory, 193
planetary motion, 55
planets
    Earth, 72–73, 78
    elliptical orbits, 81
    heliocentric motion of, 75
    Jupiter, 76, 78, 82
    laws of motion, 81
    Mars, 77, 79, 82
    Mercury, 78, 127
    motions of, 78
    musical notes of, 82 (fig.)
    Neptune, 77–78, 156
    outside our solar system, 83
    Plato, 73
    retrograde motion of, 75, 76–77
    Saturn, 78, 82
    Uranus, 77–78
    Venus, 78
Plato, 73
Platonic solid model, 79 (fig.)
platonic solids, 73, 78–80
"Playing in the Yard" (Rollins), 176
point particle (0-brane), 198
Polchinski, Joe, 197–198
poliovirus, 120, 121 (fig.)
polygons, 73

Population II stars, 155
Population III stars, 155
Porter, Cole, 17, 66
potential energy, 34
Powell, Bud, 42, 231
power spectrum, of inflation's quantum
    fluctuations, 193
preheat phase, 193
Presley, Elvis, 12, 71
"Primeval Adiabatic Perturbation in
    an Expanding Universe" (Peebles
    & Yu), 145
Principia (Newton), 102
principle of inertia, 102
The Privilege of Being a Physicist
    (Weisskopf), 8
progression (musical), 17
protogalaxies, 152, 155
Ptolemaic model, 73–74, 74 (fig.)
Ptolemy, 73–74
pure waves, 90
purpose, 215–217
Pythagoras
    creation of matter, 50
    Fourier transform, 90
    interdisciplinary focus, 3, 6
    mathematics and, 70–71
    music and, 70–72, 206
    string vibration, 72 (fig.)
    tone and, 70–71
    vibrations, 71, 101
    Western music scale, 70–71
Pythagorean followers, 67, 80
Pythagorean theorem, 70
Pythagorean theory, 166
Pythagorean-platonic philosophy, 73,
    80

quantum dynamics of the annihilation
    electron, 28 (fig.), 29
Quantum Field Theory (book), 120
quantum field theory (QFT), 7, 117,
    123, 179, 183
quantum gravity, 4, 49, 94

"Quantum Leap" (Harrison), 171
quantum mechanics, 31, 49, 162,
    175–178, 205
quantum motion, 178
quantum particles, 175–176, 176 (fig.)
quantum spin, 33, 34–38, 35 (fig.),
    37 (fig.)
quantum tunneling, 195
quasi-periodic structure, 118, 118 (fig.)

radiometers, 130, 135
radius, Schwarzschild, 138
Rakim, MC, 21–22
Ramgoolam, Sanjaye, 199–200
Ramond-Ramond charge, 198
Randall, Lisa, 198 (fig.)
Ready, Aim, Improvise! (Crook), 173
relativity theory. See general relativity
    theory; special relativity theory
Repository of Scales and Melodic Patterns
    (Lateef), 3
resonance, 113–114, 183–185,
    184 (fig.)
resonant frequencies, 148
retrograde motion of planets, 75,
    76–77
rhythm, 16
Ribot, Marc, 97
Rioux, Erin, 98, 99 (fig.)
Roach, Max, 231
Rodgers, John, 81
Rollins, Sonny, 173–174, 174 (photo),
    176, 177 (fig.)
Roxy Music (English rock band), 88
Rubin, Vera, 154
Ruff, Willie, 81
Rutherford, Ernest, 162

Sakharov, Andrei, 182–183, 185, 188
Salam, Abdus, 59, 230–231, 232
Sandole, Dennis, 65
Saturn, 78, 82
SBB (spontaneous symmetry
    breaking), 191

Schrieffer, Robert, 28
Schrödinger, Erwin, 117–118,
    167, 198
Schrödinger equation, 167
Schuller, Gunther, 173
Schwarzschild, Karl, 138
Schwarzschild radius, 138
Schwinger, Julian, 118–119
Scottish music, 224
Sen, Ashoke, 199
Shapley, Harlow, 45
Sheikh-Jabbari, Shahin, 187, 188
Siegel, Warren, 230
Simon, Paul, 88
Sinatra, Frank, 17
sine waves, 111
singularity, 138–139
sinusoidal function, 108, 109 (fig.),
    111
sixfold symmetry of whole tone scale,
    64 (fig.)
SLAC (Stanford Linear Accelerator
    Center), 185, 196
Smalls Jazz Club (New York), 41
Smolin, Lee, 86, 93–94, 139, 140, 196
soca (soul of calypso) music, 12
sonic horizon, 142 (fig.), 143–144
soul of calypso (soca) music, 12
sound, 89–90, 97, 140–142
    See also noise
sound, speed of, 141, 148, 151
sound cosmologist, 88
sound waves
    black hole analogy, 142, 143 (fig.)
    creation of, 151–152
    Fourier transform of CMB
        anisotropies and, 147 (fig.)
    gravity and, 151
    helioseismology and, 156
    musical cosmos and, 145–147
    in Perseus cluster, 143 (fig.)
    standing waves on a string, 149 (fig.)
    structure of, 154
space-time structure, 55, 86, 125, 127

special relativity theory, 65, 120, 204, 219–220, 222

speed of light, 151, 203–206

speed of sound, 141, 148, 151

Spergel, David, 145, 193, 201–202

spherical perfection, 77

spherical symmetry, 138

spin, quantum mechanics, 33, 34–38, 35 (fig.), 37 (fig.)

spontaneity, in music, 161

spontaneous symmetry breaking (SBB), 191

SQUID (superconducting quantum interference device), 31

"Stairway to Heaven" (song), 224

Stanford Linear Accelerator Center (SLAC), 185, 196

stars, 139, 154–157, 155–156 (fig.)

Steinhardt, Paul, 211

Stellar Regions (Coltrane), 5, 218

Stelle, Kellogg, 86, 206

straight ahead jazz, 95

string harmonics, 102

string theory, 6, 52–53, 55–56, 86, 196–201

string vibrations, 72 (fig.), 109–111, 110 (fig.)

strings, physics of, 104

stringy inflation models, 201

Strominger, Andy, 196

Strong City Studios (Bronx, NY), 24–25

structures, 215–217

Sun, as center of universe, 75

Sundrum, Raman, 198 (fig.)

superconducting quantum interference device (SQUID), 31

superconductor, 29–31

supergravity, 57–59

supernovas, 74–75

superstring theory, 51, 59

supreme mathematics, 21

Surely You're Joking, Mr. Feynman! (Feynman), 31

Susskind, Leonard, 139, 196–197

"Sweet Georgia Brown" (song), 224

symmetric scale, 63–66, 64 (fig.)

symmetry, 120–123, 121 (fig.), 138, 183–187, 191, 204

symmetry demonstration, 66, 66 (fig.)

synapses, 38–39

Talking Heads (band), 88

target space duality (T-duality), 52

telescopes, 45, 130, 135–136, 145

tetrahedrons (four triangles), 73

Teufel, Tim, 15

thematic improvisation, 173

theoretical physics, 117

theory of everything, 56

thermodynamics theory, 210

A Thousand and One Lessons in Supersymmetry (Gates), 229

Three Roads to Quantum Gravity, (Smolin), 86

three-dimensional structures, 47

timbre, 149–151

Timbuk 3 (hip-hop band), 24, 25

time energy uncertainty, 168–169

"A Time Varying Speed of Light as a Solution to the Cosmological Problems" (Magueijo & Albrecht), 203

Tolman, Richard, 139, 210–211

tonal center, 65, 66

tone, 16, 70–72

Topel, Spencer, 208

Townsend, Paul, 197

train levitation, magnetic, 31

Traschen, Jennie, 193

A Tribe Called Quest (hip-hop artist), 24

Tristano, Lennie, 160–161

tritone substitution, 225, 226 (fig.)

Tseytlin, Arkady, 201

Turner, Mark, 159–162, 166, 207

Turok, Neil, 211

twelve-bar blues structure, 17 (fig.)
2-branes, 86–87

U2 (band), 88
uncertainty principle, 31, 165–166, 169
universal law of gravitation, 55, 83
Universal Zulu Nation, 24
universe
    cyclic nature of, 208–210, 209 (fig.)
    Earth as center of, 72–73
    expansion hypothesis, 127–129
    heliocentrism, 76
    large-scale structure of, 28 (fig.)
    mathematical origin, 70
    origins of, 1–2
    snapshot, 146 (fig.)
    Sun as center of, 75
Unruh, Bill, 140–142, 141 (photo)
Uranus, 77–78

vacuum
    symmetry of, 183–187
    theory of, 167–168, 183
vacuum energy, 195 (fig.)
vacuum state, 181–182
vector field, 179, 180 (fig.)
vector potential, 180
velocity, 103–104
velocity graph, 105, 105 (fig.)
Venus, 78
vibration physics, 89–90, 101
Vilenkin, Alexander, 195
Vinny Idol (hip-hop producer and artist), 14
viruses, 120–122, 121 (fig.)
"Voiles" (Debussy), 63
Volkoff, George, 139
"Vortex" (Ornette), 98
vortices, 98, 200

Wally's Café Jazz Club (Boston, Massachusetts), 41

wave equation, 6
wave mechanics, 141
wave motion, 102
wave types
    cosine, 111
    gravitational, 187–188
    light, 153
    pure, 90
    sine, 111
    sound
    See sound waves
waveforms, 89–90, 111, 112 (fig.)
wavefunction, 205
wavelengths, 152
wave-particle complementarity, 167
Wayland, Sean, 225
weak nuclear force, 54–55
Weinberg, Steven, 54–55
Weisskopf, Victor, 8
Western music scale, 70–71, 223
What Is Life? (Schrödinger), 117
Wheeler, John Archibald, 126, 140
white dwarfs, 139
white noise, 190, 192
whole tone scale, 225
Wilkinson, David, 130, 132
Wilkinson Microwave Anisotropy Probe (WMAP) satellite, 145, 193, 201
Wilson, Robert Woodrow, 5, 130, 189
Witten, Edward, 51, 197, 230
Wonder, Stevie, 63

X-ray crystallography, 119

Yamaha DX7 synthesizer, 88
"You Are the Sunshine of My Life" (Wonder), 63
Young, Lester "Pres," 96
Yu, Jer, 145, 151

0-brane (point particle), 198

Lendel Marshal

Stephon Alexander is a professor of physics at Brown University and the winner of the 2013 American Physical Society Bouchet Award. He is also a jazz musician and recently finished recording his first electronic jazz album with Erin Rioux. Alexander lives in Providence, Rhode Island.